Weather or Not!

by
YORKSHIRE'S favourite TV forecasters
Paul Hudson and Bob Rust

GREAT NORTHERN

Great Northern Books
Midland Chambers, 1 Wells Road, Ilkley, LS29 9JB

The publishers acknowledge the assistance of Yorkshire Post Newspapers in providing 'Extreme Weather' photographs. Other photographs are from the collections of the authors and David Joy. Cartoons by Stephen Abbey

ISBN: 0 9544002 6 7

Printed by
The Amadeus Press, Bradford

"A meteorologist can look into a girl's eyes and tell whether or not."
– Weather forecasters' maxim

British Cataloguing in Publication Data
A catalogue for this book is available in the British Library

Contents

Paul Hudson (centre), photographed at his wedding reception in July 2003 with Bob Rust (left). On the right is Peter Levy, presenter of BBC Look North from Hull.

Introduction

The idea for this book came appropriately enough like a thunderbolt from the blue. Sheltering from the sort of storm that is only too frequent in the Yorkshire Dales, with a strong wind sweeping torrents of rain down from the fells like grape-shot, the thought occurred that Great Northern ought to add a book on weather to its list. With extreme climatic conditions occurring with increasing frequency, and the consequences of global warming becoming only too clear, it was surely a subject of concern as well as fascination.

Getting home and drying out in the early evening, a potential author came into view with equal rapidity. There on BBC Look North was Paul Hudson, presenting the forecast in his charismatic and colourful way that has become cult viewing. An approach to Paul met an enthusiastic response, with the inspired suggestion that Bob Rust, who for many years had a similar following on Yorkshire Television's Calendar, should join him as co-author.

It soon became evident that this was a writing partnership waiting to happen. At our very first meeting it was obvious that they were not only good forecasters but also good friends, going way back to a memorable evening on Calendar when Bob Rust was joined by a small twelve-year old lad from Keighley who was intent on becoming a weatherman. His name was Paul Hudson!

The five sections of this book cover distinct subject areas. The first two are on weather itself, with Bob giving a broad overview before moving on to traditional weather signs and sayings, and finally looking at weather in our region. This paves the way for Paul to evoke some of our most memorable weather extremes, ranging from storm and flood to drought and snow. The emphasis is on Yorkshire but reflects the fact that the transmission areas of both Look North and Calendar stretch down into deepest Lincolnshire and beyond.

The third section turns to weather forecasting, with Bob reviewing how this has developed from Victorian times through to the computerised world of today. The final two sections are autobiographical, with each author in turn relating how he became a weather forecaster and recalling the more memorable moments of life as a TV personality.

Conceived in a storm, this book has evolved through the glorious summer of 2003. This surely is a good omen for a work that one feels is destined to enjoy many highs and few if any lows.

David Joy
Editor

1. Our Obsession with Weather

Bob Rust

Why we are so interested in the weather

Hardly a day passes by when we do not talk about the weather. In 1758 Dr. Johnson said, 'When two Englishmen meet, their first talk is of the weather.' It is often said that we the British are obsessed with the weather, and one reply often heard is that we have weather whereas many other places have climate. It is true that in many parts of the world the weather is very predictable, but because we live on an island on the edge of a very large ocean, our weather is very changeable and often more difficult to forecast.

It could be argued that the weather affects our daily lives more than anything else. Among many other things, the weather affects what we wear, what we eat, where we go and how much gas and electricity we use. If we consider these effects on an individual we can see that the weather is very important to the country and that it can have a great effect on the economy. This is why good weather forecasts are so important and very cost effective. Even with good forecasts the weather is still our master. It can be annoying, frustrating and occasionally dangerous. It severely affects sport, holidays, communications and transport. Storms and droughts can destroy crops and livestock, and at any time the weather can threaten life on land, in the air or at sea. Man rarely feels so helpless as when confronted by the elements in all their fury.

Gas, electricity and oil consumption is greatly affected by the weather. When the weather is poor we tend to use a lot more energy. We need to keep our homes, offices and workplaces warm. We tend to have more cooked meals and use more electricity for lighting. With poor weather many people stay in their homes more than in fine weather and consume more electricity running their heating, televisions and radios. We tend to wear more and warmer clothing and this means that we have to do more washing. Drying this washing outside is often a problem and more energy is consumed with drying these clothes in dryers or in front of radiators.

Meeting these fluctuating demands for power can cause a problem for the energy producing industry. It is difficult to store electricity, so power stations in our region such as Ferrybridge, Drax and West Burton make use of weather forecasts in order to be able to meet the demand. Snow, ice and strong winds also present a problem with transmitting electricity and we can all recall how stormy winds brought about the collapse of the cooling towers at Ferrybridge in November 1965. Accumulations of snow and ice on power lines are very heavy and can actually bring down the lines. Very strong winds cause power lines to clash together and cause electrical shorting.

The weather and electricity supply are closely linked in many ways, as was dramatically demonstrated by the collapse of three cooling towers at Ferrybridge power station in November 1965. The structural failure was due to stormy winds.

Many Yorkshire folk became acutely aware of just how dramatic could be the effects of weather in 1969, when the original Emley Moor TV mast collapsed and plunged ITV viewers into darkness. The disaster was due to the weight of ice on the supporting cables.

A similar situation exists with gas. Most of our gas comes from the North Sea, and each day the gas companies make use of weather forecasts to determine how much gas will be required. Now occasionally weather forecasts do go wrong and this means that the energy suppliers may have too much or too little energy to meet the demand. In these cases there is some form of rationing or some is sold to industry a little cheaper. It is possible to liquefy gas and store it but this can be costly. However as most forecasts are reasonably correct, the power suppliers benefit economically from using forecasts.

Although very little is left of our mining industry, they are interested in weather forecasts, particularly when the pressure is falling. With just a slow fall in pressure there are no problems but when the pressure starts to fall quickly ahead of an approaching depression, this allows methane and carbon monoxide to be drawn out of the coal seams. Both of these gases are very dangerous and to overcome the problem ventilation has to be increased in order to disperse these gases.

Agriculture is also a very weather-sensitive industry. As we know all crops require rain and sunshine in order to grow. However too much or too little rain or sunshine can cause problems. Most seeds require warmth and moisture in order to germinate so an early cold spell with perhaps a frost can cause problems. With growing crops a lack of warmth or rain will affect the growth and too much rain with mature crops like potatoes, can mean that they start to rot in the ground. It can be very difficult to get farm machinery into flooded fields and crops can be lost. Dry warm weather is needed for hay making and to ripen cereal crops, but as we know a couple of heavy summer thunderstorms with their accompanying rain, hail and strong winds can cause considerable damage. Some crops can of course be dried artificially but this can be costly and affect profit margins.

Farm animals are also very weather sensitive. Grazing sheep and cattle need to have food and water brought to them when certain weather conditions prevail. A blanket of snow means the animals are unable to graze and in drought conditions the grass is soon eaten and worn away. From this it is easy to see that the weather requirements of the farmers in the Pennines and Dales, which is mostly sheep farming, differ from farms in Lincolnshire and the Vale of York where the farming is more arable and dairy. The lambing season for our Dales sheep farmers can be very difficult, with many losses if the spring turns out to be particularly cold.

Even the humble chicken can cause problems. Many are kept in poultry houses and in hot weather they can then become stressed which may lead to death. This problem is overcome by using fans to increase the ventilation. Again it is a matter of economics. The fans can be costly to run and this affects profits. Cold weather is also a problem. We have all seen chickens being transported in crates on lorries. When the weather is cold and being on moving trucks, they suffer from the wind chill effect and may die because of the low temperatures. Because of these problems, poultry producers receive forecasts to maintain good animal welfare. Within our own area these forecasts are very important as we have a number of these poultry producers, particularly in Lincolnshire and North Nottinghamshire.

With most modern farming weed and pest control is achieved by making use of chemical sprays. These are very expensive and require the right sort of weather conditions. Most sprays require several hours of dry weather to be effective and strong winds mean that the spray may be blown where it is not required, so again accurate weather forecasts are required.

Like agriculture, the construction industry is also very weather sensitive. Waterlogged sites often

(This page) *Snowfall on the North York Moors in November 1965 blocked many roads, necessitating an emergency train service to Goathland on the then closed Whitby to Pickering railway.*

(Opposite) *Serious flooding at Sheffield station in December 1981.*

make work impossible, as does a blanket of snow. Frost has a very damaging affect on mortar and concrete, and rain on drying paint can mean that the work has been wasted. Timber which is stored outside on a building site becomes saturated in prolonged rainy periods and this may well lead to problems later in the life of a building with premature wood rot. Gales and strong winds are a problem on many of the larger building sites, where tower cranes are being used, as these have very strict operating limits. Some constructors may have to make only one very large lift and hire in a tower crane for just a couple of days. The cost can be many thousands of pounds and this is wasted if the crane cannot be operated because of the strong winds. Of course roof repairers can be seen smiling after gales have occurred because this means a lot of work for them. As the saying goes, 'Every cloud has a silver lining.'

Transport is very much affected by the weather and can also prove to be very costly. We are all aware of the chaos on the roads caused by fog, ice, snow, strong winds and heavy rain. With ice and snow, the local authorities use weather forecasts to decide when and where to grit and salt. Ice often forms on roads late in the night after some rain. The salting has to be done after the rain has finished because if done earlier it would be washed away. This often means a lot of hard work in order to get the roads in a reasonable condition to meet the rush hour traffic. When large snowfalls are expected snow-ploughs must be prepared and deployed and extra manpower may need to be organised. Winter road maintenance is very well organised. Many of you may have seen little white boxes of instruments at the sides of motorways and trunk roads. These measure wind, temperature and road state. This information is fed into computers and helps with the forecasts and subsequent action by the road maintenance staff.

Weather data is also used in planning the route for roads. Before the M62 was built a lot of meteorological data was supplied to help determine which would be the best route to cross

the Pennines avoiding areas which were shown to have the worst of the weather in terms of wind, snow and fog.

Both military and civil aviation also benefits from weather forecasts. The weather conditions at both the departure and arrival airfields are important, particularly the wind, visibility and cloud base. If there has been snow or frost runways have to be treated to melt any ice. The barometric pressure is also important for the altimeter setting on an aircraft. This is because an altimeter is basically a barometer and has to be set correctly to give the true flight level of an aeroplane.

These elements of the weather are basically about safety but there are economic benefits to be gained from having good weather forecasts. At the heights at which most modern aircraft fly there are belts of very strong winds known as jet streams. In these jet streams wind speeds are generally of the order of 100 to 200 mph although as high as 350 mph has been recorded. Obviously an aircraft is very much slowed down if it has head winds of these sorts of speeds and consumes a lot of fuel. However if it has very strong tail winds it will use much less fuel. So pilots make use of weather forecasts and route the aircraft where the wind is most beneficial to the flight. In areas where there are large differences in wind speeds, like near the edge of a jet stream, we get wind shear, which causes turbulence. This as most of us know is uncomfortable for the passengers and also puts stress on an aeroplane so the pilot tries to avoid these regions.

We have two important civil airports in our region at Leeds/Bradford and Humberside. The weather factor at Humberside is very good, as it is located away from high ground and in a drier part of the region. At Leeds there are a few problems due to the altitude of the airport. Because of its height it sometimes has poor visibility caused by low cloud which is often associated with rain and weather fronts. One other problem that can occur is strong wind again caused because of the airfield's altitude. However, when we have quiet autumnal weather, it does have an advantage. In these conditions, widespread fog often forms and affects most of the lower lying parts of the area, including Humberside. However Leeds is often sat above the fog with good visibility.

Hot air balloons need to have perfect conditions to fly. In order to launch they require very light winds and at flight levels they need to know the wind patterns very accurately in order to know where they will go. Understandably they do not like flying in areas where thunderstorms are likely.

Shipping is also quite weather sensitive. The size of a vessel affects how much rough weather it is able to withstand. However generally speaking most sea-going vessels will use weather forecasts to determine the best route for their journey. They try to avoid areas of rough seas and strong winds, which are usually associated with maritime depressions, although other factors can affect the state of the sea. If there are strong winds, ferries coming into the docks at Hull are required to use tugs to ensure safe navigation of the Humber estuary and a safe docking. These have to be ordered, so the ferry captains will ask for a forecast for the Humber whilst they are out in the North Sea.

We are all very familiar with the problems that wintry weather causes our rail network. Snow, ice and flooding on tracks often means delays or cancellations, and a more recent hazard is leaves on the tracks.

Another industry that is weather sensitive is the retail trade. Wet weather is a boon to outlets for waterproof clothing and footwear, and for the sale of umbrellas, whereas with fine weather there is a large demand for garden furniture, barbecues, plants and garden tools. With fine weather there is also an increased demand for DIY equipment such as timber and paint. Sales of summer clothing, sun tan lotions and sun glasses are badly affected when the weather is unsettled but a wet spring after a cold and unsettled winter often means that travel agents sell more holidays abroad. With fine weather, cars and even houses tend to sell better and coastal retail outlets do very well, as do cafes and amusements. I always remember forecasting good weather for the East Coast on one Bank Holiday weekend. It turned out to be correct and the business people of Hornsea sent me a bottle of champagne, thanking me for influencing so many people to go to the coast for the weekend.

The type of food we eat is very much affected by the weather. It is no good stocking up with salad foods if a summer weekend is forecast to be unsettled and cool. Research has shown that we eat tinned soup in the winter, whereas in the summer we are more likely to have packet soup. Food transportation and storage are also affected by the weather, particularly when it is warm, so we use refrigeration. You can imagine what would happen if you were transporting chocolate in a non-refrigerated van on a hot summer's day.

There are many more things in our daily lives that are affected by the weather and one very important thing is the way the weather makes us feel. When the weather is fine, particularly with some sunshine, we usually feel quite happy with our lives. In periods of cold, cloudy and wet weather with little or no sunshine, we tend to feel much more down. This feeling can often mean that we are more likely to catch colds and other minor ailments and we take more time off work. In countries in more northern latitudes, where they have very little winter daylight, the suicide rate is much higher than in other parts of the world. Most of us feel better in sunny weather but it must be remembered that too much sunshine can be very dangerous and we must protect ourselves with hats, sun lotions and sunglasses.

What causes weather

Before we go any further, a few basics about the rather technical subject of what causes weather may be helpful. An envelope of gas, which is called the atmosphere, surrounds the Earth. This gas, which we term air, extends to a height of about twenty kilometres and is held to the Earth by gravity. The composition of air is very important to sustain life on Earth. The main constituents are nitrogen (about 78 per cent) and oxygen (about 21 per cent). The small amount remaining is made up of a number of gases including ozone, carbon dioxide and water vapour. The all-important constituent is water vapour, which along with other factors, causes all the weather we have on our planet. The amount of water vapour in the atmosphere is very variable, but usually no more than three per cent and in general the concentration decreases with height.

Our weather machine requires energy to drive it and we derive this energy from the sun. The area covered by a parallel beam of energy from the sun arriving at the poles is greater than the area at the equator. It easily follows that equatorial surfaces become warmer than polar surfaces. Now when most things are heated they tend to expand and this applies to the atmosphere. So the warmer surface at the equator heats the air more than does the colder polar surface. Hence we finish up with cold air at the poles and warm air at the equator. If we then consider two columns of air, one cold and one warm, the colder column is denser than the warm column and hence weighs more.

It is the weight of a column of air that we call atmospheric pressure and this is what we measure with our barometers. We finish up with high pressure in polar regions and low pressure in equatorial areas. Now nature doesn't like unequal pressures, so the heavier air at the poles giving high pressure tends to flow towards the lower pressures at the equator and this flow of air produces cells of circulation around the earth. It is this flow of air which we know as wind.

This pattern of air circulation is quite complex in itself and we have also to consider another important effect. This flow of air is taking place on a rotating Earth and as a result the winds in the northern hemisphere are deflected to the east and to the west in the southern hemisphere. This deflecting force is called the Coriolis force. As a result of these factors, we finish up with a mostly westerly flow of weather and winds in our latitudes.

So now I think we are able to imagine areas of warm and cold air being moved around the world by the winds, producing their respective high and low pressure areas. Back now to the water vapour content. It is easy to see that if the air has had a long sea track it is likely to contain more water vapour than air that has been mainly over the land as water tends to evaporate into the air. Another little bit of physics, is that the warmer the air, the more water vapour it can hold. It then follows that the warm tropical air contains more water vapour than the cold polar air.

Now where we live on this Earth is about the boundary between the warm, moist tropical air and the colder and drier polar air. It is where these different air masses meet that weather fronts are created. Fronts are depicted as red, blue and purple lines on the weather maps we see daily on our televisions, and these are the features which produce our cloud with rain and showers. Very simply the lighter, warm, moist air tends to rise over the colder, heavier air. When air rises it cools, and because it cools it is unable to hold as much water vapour. This excess vapour then condenses out as water droplets and ice crystals dependant upon the temperature. This condensation usually takes place on particles of soot, dust or salt, and these are known as condensation nuclei.

It is these masses of water droplets and ice crystals which we see most days and know as clouds. As these clouds get bigger, the droplets and crystals become larger and more abundant and eventually fall from the clouds as rain and snow.

The above broad-brush explanation relates to the large-scale weather systems that develop, move around the Earth with their weather and eventually die. We do have a lot of smaller weather happenings and these again rely on the same physical principles.

Let us now take a look at some of these smaller weather events. If we have the wind blowing from the west, it then arrives at the Pennines. Now some of the air will blow through gaps in the hills but most of the air will be forced to rise and go over the tops. As we have seen earlier, the rising air is cooled and this in turn causes condensation, which forms clouds that are eventually likely to produce showers. As the air comes down the eastern side of the Pennies the process is reversed as the air warms, condensation ceases and the clouds dissipate. This is a common happening that we have all heard on weather forecasts as 'Showers over the Pennines'.

In the winter we often have frost and fog on the low ground, particularly in the Vale of York. This usually happens on a cold clear night with very light winds. Temperatures fall because the earth is radiating away the heat it has obtained from the sun during the day. I say a clear night, because if it is cloudy, this cloud acts like a blanket and reduces the heat loss. Now the cooling air on the top of the hills becomes heavier and tends to roll down the hills and accumulate in the valleys. These large areas of cold air in the valleys cannot hold their water vapour, which condenses out as water droplets. As the wind is very light this suspension of droplets is dispersed through the lowest couple of hundred feet producing fog. If the wind is absolutely calm, these water droplets are just deposited on the ground producing a heavy dew.

Should the wind increase after the formation of fog, this will tend to disperse the fog or lift it onto higher ground. The accumulation of cold air in the valleys tends to increase the cooling making frost quite likely.

In summer as the sun is shining down the land is warmed. The sun is also shining on the surface of the sea, but the sea is turning over and moving, so the heating effect is spread downwards through the sea. This coupled with the difference in nature between the land and the sea means that the air over the land becomes warmer than the air over the sea. This temperature difference leads to pressure over the land being lower than over the sea and air then flows from the sea to the land, producing sea breezes with their consequent cooling effects.

Another common situation regularly occurs during the spring or summer. We often wake to a sunny morning with cloudless skies. With this promising-looking weather we decide to have a day out. By the time we have loaded the car and set out on our journey we find that cloud is already starting to bubble up and frequently by the time we reach our destination there are showers developing. What has happened here is that early in the morning the ground is cool although the sun is shining. With more sunshine during the morning, the ground temperature starts to rise and this in turn starts to warm the air. If we think of a parcel of air being warmed it begins to expand and become lighter. This parcel of lighter air will then start to rise. We have seen that rising air cools and cannot hold as much water vapour, so condensation takes place in our parcel of air, with clouds developing which often give showers. If these clouds become big enough they can also produce thunderstorms. Come the evening however, temperatures start to fall and the clouds dissipate to give a fine evening and generally a clear night.

Now we have discussed some of the physical effects that cause weather and some of these we are able to see every day in and around the home. The shimmering effect we see above radiators is due to the expanding and rising of air being warmed by the radiator. If we take a look in the refrigerator or freezer, we can see water droplets and ice deposits. These are caused by the air being cooled and then no longer being able to hold its water vapour, so we get condensation with ice and water droplets. A similar effect can bee seen on windows, particularly those not double-glazed. Cold outside temperatures cools the glass and this in turn cools the warm air in the room, giving condensation on the glass. This effect is enhanced, particularly in the kitchen, where cooking is likely to be putting more water vapour into the air.

These are just a few of the many effects a forecaster has to consider when making a weather forecast and they will be discussed later in the book.

Changing weather patterns

In this country we are all very aware of the changing weather. It is often quite dramatic from one day to another. However in our latitudes this is quite normal and is what we should expect due to our maritime climate.

It is the more significant changes that we are interested in, when the weather patterns start to change more dramatically. We all know what kinds of weather we should expect in our four seasons. When these change markedly, and the changes persist for several years, this is an indication that something is going wrong in the expected weather pattern. It is these changes that we call climate changes and are usually happening throughout the world. When climate change is happening, it usually means that weather records are frequently being broken.

Intense periods of rain and consequent flash flooding are becoming more frequent. A young butcher has an uphill struggle in trying to divert water away from shop cellar in December 1983. The location is thought to be Otley.

Normally when the weather is behaving as expected, the yearly totals of weather elements such as rainfall, sunshine, frost days are close to the long-term average. Such averages would generally be over a period of say a hundred years.

We know that there have been significant changes in the climate in the past, such as the Ice Age when ice spread as far south as the Great Lakes in America and Switzerland in Europe. Although data is somewhat sparse and not totally reliable it would appear that there were as many as twenty occasions when the ice and glaciers spread south. In between there were warmer periods called Interglacial stages and during these the climate was actually warmer that it is now. Apart from this Ice Age it is known that there have been several more, probably about six. The reasons for these great climate changes are quite complex, but it may simply have been changes in the Earth's orbit around the sun or a slight tilting of the Earth's axis. Both of these effects would mean a large change of heat energy from the sun arriving at the Earth, leading to global warming and cooling. These effects may well be part of a natural cycle over millions of years but because of lack of data it is impossible to predict if or when they will occur again in the future.

About forty years ago the scientists were predicting that the Earth was cooling and we were probably heading towards another mini Ice Age. However over recent years there has been a complete turnabout, and we are all familiar with the notion that global warming is now happening and is leading to significant changes in the Earth's weather patterns. So what is global warming? Our planet is designed so that each day it receives it's warming from the sun in the form of short wave radiation. This warming of the Earth then warms the atmosphere and drives the weather, producing the climate we are used to. The atmosphere also absorbs some heat as the sun's energy passes through it and at the same time the ozone absorbs the more dangerous radiation that can cause skin cancer. As the Earth is receiving heat from the sun it also loses heat in the form of long wave radiation. However all this heat is not lost to space because we have the atmosphere. In the atmosphere we have clouds and water vapour and some other gases, mostly carbon dioxide, which trap and absorb some of the heat and even reradiate it back down to the Earth and this prevents the Earth cooling too much.

Every thing is very finely balanced so that we have a climate where man can survive quite easily, with predictable and not too severe weather patterns. Each day, everywhere on Earth receives warming from the sun and overnight there is some cooling and this remains reasonably constant leaving our atmosphere at a fairly steady temperature giving us the familiar world weather patterns.

It is this state of heat exchange balance that is known as the Greenhouse Effect. We can understand this better by thinking about our own garden greenhouses. The heat from the sun is able to pas through the glass into the greenhouse, warming the ground and the air. However the heat radiation from inside the greenhouse cannot easily pass through the glass into the open so the temperature inside is kept high.

Now it is when the finely balanced heat exchange is altered that we get global warming and the consequent changes in our expected weather patterns. We could produce a similar effect in the greenhouse by using double-glazing. This would reduce even further the heat loss from inside the greenhouse and produce even higher temperatures.

On the moon where there is no atmosphere the temperature soars to 110 degrees Celsius in the sun and falls to minus 170 degrees Celsius in the lunar night. On the planet Venus there is an atmosphere almost entirely composed of carbon dioxide. This causes the temperature near the surface to be around degrees 428 degrees Celsius.

This is hot enough to melt lead and about twice as hot as the average kitchen oven. Unlike on our Earth, man could not survive in these climates.

So it can be seen that the amount of carbon dioxide in the atmosphere can have a serious effect on the climate. It is the increase in carbon dioxide and other gases that has led to our present concern about the changing climate we appear to be experiencing. This increase in these greenhouse gases has been taking place over a long period throughout much of the developed world. For many years we have been burning more and more fossilised fuel such as coal and oil and consequently producing more carbon dioxide. One other important factor is that trees and plants are able to absorb carbon dioxide. Sadly at the same time that we have increased the amount of carbon dioxide we produce, we have been clearing the rain forests which would have helped to remove the gas. The result is that the atmosphere has been getting warmer and altering the expected weather patterns.

Probably the thing that most of us have noticed is the change in the winter weather. The amount of cold weather with frost and snow is now considerably less than we would have expected. In the past there have been mild winters but the present trend is rather worrying. Apart from the lack of cold, wintry weather, we do seem to be having more periods of rain, which has led to some serious flooding. When we have even large amounts of snow, thawing is usually fairly slow and the river network can cope with the extra water. However with long periods of rain the run off is immediate and the rivers rise and often flood. Many of our drainage systems are designed to cope with normal rainfall, but with the large rainfall amounts we seem to be having at the moment, they are unable to cope and this again leads to flooding. Economic pressure and the demand for more homes have led to building on flood planes which has proved to be a bad decision. Some river courses have been altered and there does not seem to be as much dredging as there used to be. All of these factors have contributed to the flooding and the problems that it has caused.

From all of this we can see that man has not only contributed to the cause of the changing weather but has also not helped with the problem once it has arisen. Although perhaps not as marked as the changes in the winter weather, the other seasons have also changed somewhat. We have had most of our best summers in the past thirty to forty years and the seasons seem to have merged into each other, with some very variable springs and autumns.

What may happen to the weather in the future is somewhat difficult to determine. Many forecasts have been made using mathematical models and many make quite different predictions. Other factors, such as melting of the ice caps as a result of the warming atmosphere, will affect the weather in the future. It is the presence of the cold seas in the Arctic that contribute to the Gulf Stream effect that keeps our climate mild. The eventual outcome is in doubt, and opinion among eminent scientists does differ. There are still some that believe what we are experiencing is just a natural blip and will eventually go away and be forgotten.

Whatever the future, the warning has been worthwhile. We are all now more aware of the damage we may have done to our environment and many steps have been taken to try and correct the problem. We use much less fossilised fuel and it is burnt in a more efficient way. We use more and better insulation in our buildings, reducing heat loss and so using less power. The motor car is now considerably more efficient producing much less pollution, we are planting more trees and trying hard with most things to be environmentally more efficient. Obviously there are pockets of resistance because the economic strength of many countries is dependent upon

One consequence of the apparent global warming is that harsh winters with heavy snowfalls are becoming much less common. This picture was originally captioned, 'No school -- snow school?' Depicting a Middlesmoor schoolgirl, it may well have captured a scene that will rarely if ever be seen again.

pollution-producing industries. The occurrence of smoke-laden fogs is now a thing of the past, which means we are all healthier and reduces demands on the health services. One other very important thing is that the younger generation is now very much aware of environmental problems and hopefully will not make the same mistakes as past generations have done. So provided the whole world tries to co-operate and make improvements the world should become a better place to live in.

Although we are able to take some action to improve our present problem with global warming, there are some things that are beyond our control. One of these is volcanic eruption. An erupting volcano throws up a considerable amount of ash and debris into the atmosphere. This ash then circulates around the globe on the high level winds and affects the amount of radiation arriving from the sun. Weather patterns in many parts of the world are likely to be affected but the effect should be short lasting, as the ash becomes more dispersed or even falls back to earth.

The warning of global warming has inspired the need to seek out new sources of energy. Nuclear power stations are one way of meeting our energy needs but there is a lot of concern about their safety. They are costly to maintain and to shut down at the end of their lives. A lot of work is being done to improve their safety but we do not want to see another Chernobyl. Solar power is also being used in a small way but it does have some problems. Solar cells need to be large to make it worthwhile and the lack of sunshine in cloudier climates obviously limits their use.

Harnessing the energy from waves and from the wind is now being tried and is quite successful although amounts produced are still quite small. The motion of the wind and the waves is used to turn turbines and these drive generators to produce electricity. Here in the North of England we already have some successful wind farms on the Pennines and some are being planned for Lincolnshire. It would seem on the face of it that using wind and wave energy is a sensible way forward. However one has to ask how many of these wind farms and wave machines we might need to make a really worthwhile contribution and the answer is probably a very great number. Now this in itself may cause us another problem. Waves and wind are interconnected and are all part of our weather and atmosphere. In Yorkshire we would say, 'Yer don't get owt for nowt', and Isaac Newton would be a bit more scientific and say, 'Energy can neither be created nor destroyed.' This means we would be extracting our energy needs from the atmosphere, and would this in time affect our weather patterns as global warming seems to have done? We are obviously a long way from that position and it may mean that the whole of Europe would have to be covered in windmills to have any effect on the weather.

Traditional weather signs and sayings

Nowadays most of us make use of the usually very good weather forecasts produced by the Meteorological Office to plan our day's activities. Now scientifically produced weather forecasts were not available until about 150 years ago but people were still very interested in trying to predict the weather. They were very aware of changes in the weather and many advances had been made into producing instruments that could measure some of the different elements of the weather.

Galileo can probably be credited with producing the first thermometer in 1593 and he called this instrument a thermoscope. However it took until 1641 to produce the first thermometer similar to those we have today. This was called the Florentine thermometer and was made by Ferdinand 11, the Grand Duke of Tuscany. It used alcohol in a sealed glass tube. Many other thermometers followed mostly using mercury instead of alcohol. All of these thermometers had different scales and it was Gabriel Fahrenheit who in 1714 consolidated all these different ideas into the Fahrenheit scale with which many of us are familiar. The unit degree was small enough to record temperatures with adequate precision in whole degrees and, with 32 degrees being the freezing point of water, there was less need to use a negative sign which can be a source of error when printing.

Anders Celsius produced his temperature scale in 1742 where there are 100 degrees between the freezing point and boiling point of water. This was widely adopted by the scientific world because of the simplicity of the metric system and is mostly used today. It is worth noting that on the original Celsius scale, 100 degrees was the freezing point and 0 degrees the boiling point but this was sensibly reversed to the present-day scale.

Evangelista Torricelli, who was a pupil of Galileo, can be credited with inventing the first barometer. As with the thermometer there were many improvements made over the years leading to the mercury barometers we see today. Most household barometers nowadays are aneroid, which means the rather dangerous liquid mercury is not necessary. These have other advantages in that they are easy to transport and they are accurate enough for most practical purposes.

Now although there were instruments around for measuring most elements of the weather for a long time, no one was able to produce a weather forecast. It was often preached and thought that God sent bad weather as some form of punishment and people prayed in church for good weather to ripen crops and bring a good harvest.

An interesting extract from the diaries of Bilton Church shows this.

April 6th to July 12th 1826

During this summer there was experienced a longer continuance of drought and more excessive heat than had been remembered of many years. Little rain having fallen for upwards of around three months, in consequence of which all the ponds in the neighbourhood were dried up. The pastures became bare and withered beyond example and very little hay, or oat or barley straw obtained for winter fodder. The wheat however was abundant, and owing to the favourable autumn and open winter which followed, there was not the dreadful scarcity which according to all human probability might have been anticipated. Thus does an all wise Providence temper his chastisement with mercies, and did all things ultimately for our good.

However, people started to look around them to see if they could detect signs in nature as to what was happening to the weather. This was mostly done by people who worked outdoors, such as farmers, and led to what we now know as weather lore. Much weather lore is all but useless and little more than superstition but based on close observation of the natural world can give some clues as to what might be happening to the weather. Small variations in the weather that we do not necessarily notice can be detected by plants and animals, which may change their appearance or behaviour.

So firstly let us take a look at a few of these weather sayings that do have some scientific basis and are occasionally true.

Many people believe that the moon has strong connections with the weather and can be used in predicting what the weather may do. The moon is very useful for observing the sky at night and occasionally is surrounded by a halo. This halo is cloud made up of ice crystals which is often a forerunner of rain. In the past the language was much more poetic than nowadays and people wrote things like, *The moon in haloes hides her head,* or *If she should clasp a dark mist within her unclear crescent, heavy rain is in store for farmer and fisherman.* So though not completely reliable, a halo around the moon may well mean that rain is on the way.

When the moon is sharply visible, it means that there is little or no cloud around. This would lead to rapid cooling overnight and in winter frost is always likely. So the saying, *Full moon, frost soon,* has some scientific reasoning and is often correct. A more updated version that is just as reliable is *Clear moon, frost soon.* Even today a lot of people think that a change in the phase of the moon will bring about a change in the weather. This is totally untrue and brings me on to the best rhyme about the moon that I know. It is:
*The moon and the weather **may** change together,*
*But a change in the moon **will not** change the weather.*

Not so often heard is:
If there is frost or dew, or morning fog
No rain this day will you log.
Here frost, dew or fog is most likely with settled weather conditions. Usually pressure is high and rain unlikely. So this is a fairly reliable saying.

The state of the sky is a good indication of what the weather may do and is used by forecasters today to help with their forecasts. Approaching weather fronts always give thin high cloud at first that gradually lowers and thickens, eventually bringing rain. It is this which led to probably one of the best-known rhymes in our weather lore:
Red sky at night, Shepherd's delight.
Red sky in the morning, Shepherd's warning.
This one has a scientific basis and is quite often true. The explanation is that with a red sky at night, the sun setting in the west shines on decaying cloud in the east. As we know most of our weather travels from west to east and the decaying cloud has probably past over us giving poor weather with rain. Now the usual pattern is for one poor day and then a fine day before more weather systems move towards us from the west. So the setting sun is shining on the weather that has past, so the chances are that the next day will be fine. Now with a red sky in the morning, the sun rising in the east shines on the underside of high cloud out to the west. High cloud is often the forerunner of rain, so rain can be expected later.

Another saying that is occasionally true is:
Rain before seven, Fine before eleven.
This is based on the fact that the average period of cloud and rain associated with a warm front moving from the west is about four hours.

Wind from the east, For a fortnight at least.
There again is some truth in this. Often when we have an easterly wind, it is caused by a blocking high-pressure system over Scandinavia. This is a fairly common winter situation and can lead to a prolonged period of cold weather.

Whilst talking about wintry weather, it is often said when we have had a good fall of snow, and it has not thawed, that it will take another lot of snow to move it. There is very little truth in this except that it usually requires warm air from the west to melt the snow. We normally get this warm air from an Atlantic weather front that will also bring rain. As this rain approaches the snow fields, it is likely to start falling as snow before it turns to rain and everything gets milder giving a thaw. Another popular belief is that it is too cold to snow. This has an element of truth in that we usually get our coldest weather with clear skies and hence no snow. However things can and do change and a weather system may well bring the cloud and snow. What we usually find is that the temperature usually rises to about zero degrees Celsius in snow.

The barometer can give a very good indication of what the weather may do. Basically the best advice is to ignore the words written around the barometer. It is how the pressure is changing which is important. So a daily check will reveal

The wind features in many traditional weather sayings, such as 'Wind from the east, for a fortnight at least.' These stunted trees on Fylingdales Moor will have seen many a 'lazy wind', when it blows through you instead of around you.

how the pressure is changing. This is why there are two pointers on a barometer. One is the needle which indicates the pressure and the other, which you are able to turn, is there as a memory aide. Line them up together and the next day you will be able to see how the pressure has changed. If the pressure is rising the weather is likely to become more settled, but with falling pressure the weather should become more unsettled. There are a number of rhymes associated with the barometer and many are reasonably sound:
Sharp rise after low, Oft foretells a stronger blow.

A greying sky and falling glass, Soundly sleeps the careless ass.

Most of us have heard the saying,
Ne'er cast a clout, 'til May is out.
Clout refers to clothing. May refers to the hawthorn bush, not the month, although the May blossom is usually out in the month of May. This is fairly reliable, as statistics show that after the end of May, frost is highly unlikely. It is probably the best time to plant out bedding plants, as any earlier and the frost may take them.

There are many signs in nature which are associated with weather changes, but I believe these are just reactions to what the weather is doing and not what the weather will do, although my old friend Bill Foggit made himself famous by using these signs in nature.

Probably the best known of these signs is changes in the appearance of seaweed and cones. With seaweed, people who live near the coast often hang it outside. In fine weather it shrivels and is dry to touch, but when rain threatens it swells and feels damp. If this does happen it is probably because the seaweed reacts to extra moisture in the atmosphere but it does not necessarily mean that it will rain. Similarly with pine and fir cones. Their scales shrivel and open up when the weather is dry but when the air is moist, the scales absorb the moisture, making them pliable, and allowing the cone to regain its normal shape. Again this does not mean that it will rain. With both the changes in the cone and the seaweed, it is more a detection of more moisture in the air and not necessarily a sign that rain is on its way.

Human hair also reacts to a change in the moisture content of the air. When the air is dry it

Whether this torrential rain in Leeds in June 1971 is 'teeming', 'peshing', 'chucking', 'tippling' or 'siling' it down is an open question. But the mere fact that there are so many terms for rain 'up north' does suggest that we have quite a lot of it!

tends to shrink and curl. Professional weather observers make use of this effect to measure the humidity of the air. Strands of human hair are used in an instrument and the changes in length of the hair are converted into a humidity reading and recorded on a graph. Many flowers react to changes in the weather and they are again just an indicator of what the weather is doing and will not foretell future events. A couple of well-known flowers which react to changes are the Scarlet Pimpernell and Morning Glory. With both of these the flowers open in fine weather and close when rain is around.

When hollow winds begin to blow, The clouds look black, the glass is low. Closed is the little pimpernel.

Although the professional weather forecasters dismiss most or all of these signs and sayings, they were the only things available before we had weather forecasts. In a way it is much more romantic to have Bill Foggitt forecast a good summer because of the behaviour of moles than Paul Hudson to give a scientifically based forecast for the next few days. Whether Bill's forecast turns out to be correct or not does not really matter at the end of the day.

Many of the old ideas were written in rhyme, presumably so that they could easily be remembered. There are literally hundreds of these around but space allows for only a few here:

Onion skins very thin, Mild the winter coming in.
Onion skins thick and tough, Coming winter cold and rough.

When a cow tries to scratch its ear, It means a shower is very near.
When it thumps its ribs with its tail, Look out for thunder, lightning and hail.

St. Swithin's Day if ye do rain, For forty days it will remain.
St. Swithin's Day an' ye be fair, For forty days 'twill rain nae mair.

This is based on one occasion only. On 15 July 971, the body of Swithin, who had been the Bishop of Winchester, was due to be reburied in the cathedral. Since his death in 862 he had been buried in the churchyard at his own request, where he wanted the rain to fall on his grave. On the day, the rain was so heavy it had to be postponed. The wet spell continued for perhaps forty days and this was taken as a sign that Swithin's last request to be buried in the churchyard should be respected, so there he was left.

One relating to trees coming out in leaves is
The Ash before the oak, You're in for a soak.
Oak before ash, Ther'll be just a splash.

Even spiders and poultry have their place in weather lore,
When spiders from their holes do creep.
The skies above are sure to weep.

If the cock goes crowing to his bed,
He'll certainly rise with a watery head.

Ice in November to bear a duck,
There'll be nowt after that but sludge and muck.

Some of these sayings are quite amusing and a couple of old Yorkshire ones I like are:
The faster the rain, the quicker it stops.

Courtin''ll cease when t' garse is out o' flower.
(I think this means that when there are no flowers on the gorse, it's too cold for outside courting).

It is often the elderly farmers and people who have worked outside that have their own ideas and sayings about the weather. One story goes that when you asked one old Dales farmer what the weather might do, he would look up at the sky, glance round at his stock, point to a passing rook, and say, 'Aye, it could do owt.'

Another common one that is heard in many places in the region is, *If you can't see so-and-so* (where this would be a local landmark such as a church or a hill), *then it's raining. If you can see it, then it will soon be raining.*

In many parts of the region, February is referred to as 'February Filldyke'. This is quite surprising as February is on average one of the driest months of the year. It is probably called this because of snow melt, with the snow probably having fallen during January.

Some years ago in America, several hundred of these sayings were investigated. They came up with the conclusion that 15 per cent were wrong, 17 per cent could mean anything, but over 60 per cent were reliable. The rest were undecided. I think that I prefer the Dales farmer's conclusion, 'It could do owt.'

Across our region we do have different local words for describing the weather. When it is cold people say that they are 'nithered' or it's a 'nithering wind'. Strong, cold winds are often described as 'lazy winds' followed by the explanation that they blow through you instead of around you. If you are cold, people over many parts of the region say that they are 'starved'.

We have several ways of describing rain intensity, some of which can't be discussed here. Among these are 'teeming', 'peshing', 'chucking', 'tippling' and 'siling'. The origin of 'siling' is quite interesting. I remember one occasion on television, I asked if any viewers knew how the expression 'siling it down' had originated. The response was quite remarkable. I received about two hundred letters from viewers giving various explanations about the origin of the word. The explanations were extremely varied but the majority were correct in that it originated from the word 'sile'. This word is now out of use but was a piece of equipment used by farmers. When cows were being hand-milked, bits of straw used to get kicked up into the milk. The farmer then used a crude filter to clean the milk and this was called a sile. Being just a course filter, the milk would rush through quickly, similar to heavy rain. Although I got the correct answer, it made me a lot of work in that I had to answer each one of the letters.

One description of the weather that is not very often heard is 'stowering'. This is an expression used mainly in the old North Riding, when a blizzard or blowing snow makes it difficult to see ahead.

Many people claim that they have pain when it is going to rain. Some say that they get trouble with different joints, which is a sign of rain. Others claim are that their corns hurting is a sign of coming rain. I suppose that it is likely that some joints can ache when the moisture in the air is increasing. However I don't think that these events are a reliable indicator of coming rain. Having said that, I will always remember the Calendar presenter Roger Greenwood. He would tell me that his knee was painful when rain was expected and he was invariably correct.

To conclude this discussion about weather lore and local sayings, I am reminded of an old gentleman from Kettlewell who, when greeted on a stormy morning with 'Rotten weather we're having Jimmy', replied, 'What 'll you say if you wakened sum day an' ther wor noa weather?'

Our weather in the region

"Did you lock the door?"

I have often heard it said that Harrogate is three overcoats colder than Wetherby and on a number of occasions I have found this to be so. Now Wetherby and Harrogate are only a few miles apart and we get big variations in the weather between the two places. If we consider the whole of our region there are quite immense differences in the weather pattern from place to place, and these are mainly due to the topography.

So let us take a look at our region. It covers a total area of about 21,325 square kilometres. It has a coastline of about 250 kilometres stretching from the estuary of the Tees to the Wash. At its widest

The Pennines have always been susceptible to the effects of the prevailing westerly wind. Brimham Rocks dramatically display the results of countless centuries of wind erosion.

point the region stretches to about 160 kilometres and comes within 15 kilometres of the West Coast at Morecambe Bay. The western border is the Pennines running down into the Peak District, with the highest point being Whernside on the Cumbrian border at 737 metres. Moving east from the Pennines we have the lowlands of Lincolnshire and the Vale of York. Towards the East Coast we have the North York Moors, and the somewhat lower yet still important Yorkshire and Lincolnshire Wolds. This is all a very mixed topography and greatly contributes to our very mixed regional weather.

Let us firstly look at the weather in the Pennines and the towns and cities on the foothills such as Leeds, Bradford, Skipton and Sheffield. Now all of these places are high. As you go up into the atmosphere the temperatures fall so it is reasonably fair to say that these places are colder than low lying parts. Having said that, we will often see that places like Leeds can record very high temperatures in the summer. This is because they are like heat islands, all the brick and concrete buildings absorbing and re-radiating the heat and pushing up temperatures. This heat island effect also benefits the town and cities overnight, when temperatures are kept somewhat higher than nearby rural areas as the buildings and concrete act like a storage heater.

In a similar situation we will find that temperatures at Leeds/Bradford airport are lower and more representative. Although Pennine areas are generally colder than elsewhere, there are occasions overnight when they are less cold than some other places. When conditions are clear and calm the cold air tends to roll down into the valleys, where it accumulates and forms frost hollows. The largest and most significant of these is the Vale of York.

As most of our weather comes from the west, it is Pennine areas that are first affected. A weather front with its accompanying rain arrives at the Pennines and as it is forced to rise over the high ground it is given extra impetus and the rain intensity is often increased. In a wintry situation the precipitation is likely to be of sleet or snow depending upon the temperature. This explains why forecasts often say rain or snow, 'heaviest over the Pennines'. Even in the Pennines or Peak District, precipitation amounts will vary a little, with the heaviest falls being on the highest and windward facing parts. In a showery situation it is again the high ground which is likely to get most of the showers and these may well be accompanied by thunder and perhaps hail.

Apart from being the wettest parts of the region, the Pennines are also the cloudiest. Air rising over the high ground will produce cloud, and even a cloud sheet from the west may be broken as it moves to the east of the hills. The situation described here is generally when we have westerly winds. When we have winds from the east the cloud may well break to make Pennine parts sunnier than the east. In the rather cloudy and wet picture we paint of the western high ground, the cloud is often low and shrouds the hills giving fairly widespread hill fog.

We have now only to talk about how windy are the Pennine areas. As we go up into the atmosphere the winds usually increase with height. This is because they are away from the frictional effects of the ground, which tends to reduce wind speeds. It then follows that the windiest parts of our region are generally the Pennines and the Peak District, although there are occasions when this is not the case and these will be discussed later. So in summary, the western high ground is the wettest, coldest, cloudiest and windiest part of the region.

Moving away from the high ground in the west, much of the rest of the region is low ground as far as the coast. We do however have some high grounds, these being the North York Moors and the Wolds of Yorkshire and Lincolnshire. There are a few other hills, but their effect on the weather is not so great, although to the locals they may seem rather important. Now like the Pennines the North York Moors and the Wolds do have their own weather which differs somewhat from the weather over the other low ground to the east of the Pennines.

As we described for the Pennines and Peak District, the Wolds and North York moors have more rain than the surrounding lower parts. Their annual rainfall figures are not as high as the Pennines, although the North York Moors, being the highest part of the eastern high ground, are appreciably wetter than the other eastern parts of the region. Like the Pennines, in wintry situations considerable snowfall can occur in these areas. Snow can become a real problem in situations where we have an easterly wind. These occasions are likely throughout the winter and can continue into April. What happens is that very cold air from the continent travels over the North Sea, which is relatively warmer. This warmth enhances the activity of weak weather fronts or sets off showers, which can put down a lot of snow in quite a short time. As these showers move westwards they tend to die out over the colder land.

Not only are the Wolds and North York Moors wetter than other eastern parts, they are also cloudier. This cloud is often low if accompanied by rain, and like the Pennines they will become misty with fog at times, particularly over the North York Moors. Because of their altitude, temperatures on these eastern higher parts are a little lower than the surrounding low ground. As in Pennine areas, on still clear nights the heavy cold air will drain into surrounding valleys, creating frost hollows, a notable one being the Vale of Pickering. Like the Pennines, the Wolds and Moors tend to be windier than the lower parts but again there are some marked exceptions, which we will discuss later.

So to summarise the weather affecting the Wolds and North York Moors, it can be fairly said that it is not as pleasant as the surrounding low ground but it is generally better than most Pennine areas.

Considerable snowfall can occur in the Yorkshire Wolds, especially on an easterly wind bringing cold air in from the continent.

(Top) *Postman Ernest Fleetham has a long trudge to deliver mail to Gritts Farm, Fridaythorpe, in February 1954.*
(Lower) *The main York to Bridlington road is utterly impassable at the summit of Garrowby Hill in January 1962.*

Classic combination at Scarborough, with high tide and a north-easterly gale giving cars on the Marine Drive a salty wash.

Now we are left with the remainder of the region, which is virtually all low-ground. It comprises the Vale of York, stretching down into the lower parts of South Yorkshire into Lincolnshire down to the Wash and taking in the Trent Valley. These areas really have the best of the weather.

The rainfall is lower than all other parts with the driest spots being in South Lincolnshire towards the coast. The only places in the country drier than South Lincolnshire are parts of East Anglia and North Kent. The reason is that as most rain-bearing weather systems come from the west, they are considerably weaker by the time that they arrive in these areas. There are of course some exceptions, and one is the Trent Valley. Here they have the highest incidence of thunderstorms in the British Isles, with over twenty days a year on average. Elsewhere in the region the average is around five to ten days. The reason for the high frequency of thunderstorms in the Trent Valley is the number of power stations in the area. They act as a local source of heat and moisture, which trigger off their own cloud systems. We have all seen this effect when driving along our region's motorways. The sky is often clear of clouds but above the power stations there will be some cloud bubbling up. In a showery situation, which is usually with winds from the west or north-west, most places are protected from the showers by the Pennines. However where there are sizeable gaps in the hills some showers do filter through.

Not only is the low ground the driest part of the region but it also has less cloud and consequently more sunshine. The reason is much as that explained for the lower rainfall amounts. Not only do the weakening weather fronts produce less rain, their associated cloud tends to break up after leaving the hills.

These lowland areas usually have the highest temperatures. Boston is often one of the warmest places in the country in a summer situation, but some of the region's cities will come up with quite high temperatures because of the buildings effect

The Bridlington lifeboat sets out to escort fishing boats to safety. A wind with an easterly component always produces rough seas on the East Coast, even though the actual wind speed may not be very strong.

as I have mentioned earlier. Although daytime temperatures are higher, overnight they are usually lower than the higher ground. The reason, as explained earlier, is because the cold air on the hills sinks down and accumulates over the low ground. This means that in winter some of these areas have the highest incidence of frost in the region.

As well as having most frosts, fog is also a problem over lower ground. Fog is caused when air is cooled enough for water droplets to form, and often accompanies the low temperatures and frosts over the low ground. Probably the worst area for fog is the Vale of York. Here fog forms and often persists throughout the day when elsewhere it is likely to be sunny. On these days when the fog persists, temperatures will be well below those in the sunnier areas. Although I cannot offer a reason, it is worth noting that fog does not generally persist throughout two days. It is often forecast to persist throughout the second night and day, but there are numerous occasions when it has cleared by the second morning.

One other interesting fact is that many places do not get fog if the direction of the wind is from the north-west. Conditions for fog formation may be perfect with clear skies expected and a very light wind. All the meteorological calculations would lead to the conclusion that there will be fog, but if the drift is from the north-west, fog will not form.

Looking now at the wind pattern over the lower parts of the region, it is fair to say that on most occasions the high ground protects the area from the strongest winds. Like everything else, there are some exceptions. We have already mentioned how showers can filter through gaps in the hills and similarly wind funnels through these gaps to produce some strong winds to the east of the Pennines. One notorious gap in the Pennines is the Aire valley, which can give very strong winds in the Leeds area. Other Pennine valleys have a similar effect over other parts of the region.

There is another situation that can produce some very strong winds away from the Pennines.

Occasionally the atmosphere is in such a state that as it crosses the Pennines it starts to oscillate and is then in the form of waves across the area. This situation can often be recognised by the presence of almond shaped clouds, which do not move despite the wind. What happens is that in the troughs of this waving atmosphere, the strong winds flowing down to the ground reinforce the natural low-level wind producing exceptionally strong surface winds. These strong winds are likely to persist throughout the day and only a few miles away the wind will be much lighter. It was this sort of situation that caused the notorious Sheffield gales in February 1962.

However, overall one easily concludes that the weather over the lower parts of the region is really much better than over the high ground.

Now this general description of the weather applies to most of the low ground but the coastal strip itself deserves a special mention. The important factor here is the sea, which has quite an effect on the weather. The temperature of the sea along the coast is very important. Over land we can see temperatures varying between 30 degrees Celsius at the height of summer to as low or lower than minus 10 degrees in winter. Over the coastal waters off the East Coast the variation is much less. In summer, sea temperatures are unlikely to rise above 14 degrees Celsius falling to no lower than 7 degrees in the winter.

The first problem the coastal strip has is the frequent formation of low cloud or sea fog often called sea fret or 'haar'. This is most likely during late spring and early summer but can happen at other times. If the wind is from the east and the air is moist and warm, the sea cools down the air causing condensation that forms fog or low cloud. This fog and cloud will move inland somewhat but will soon burn off with the higher land temperatures. In these conditions the coast itself will be much colder than inland places. The cold seas also cause sea breezes which keep the coast cooler than inland. High inland temperature causes the air to rise, which is then replaced by the cold air from the sea causing temperatures to fall along the coast. There are days when temperatures inland can be as much as 15 degrees warmer than on the coast. In winter, when there is a drift from the sea, the temperatures overnight will be much higher than inland making frost much less frequent than inland areas.

East or north-easterly gales do affect the coast but are reduced in speed over inland areas, and these same winds are likely to bring sleet and snow showers to the coast during the winter. For those who sail in the coastal waters of the east coast, the main problems are sea fog and the sea conditions. Whenever the wind has an on shore component it will produce rough seas even though the wind speed may not be very strong. With strong or gale force northerly winds the situation becomes very serious. Considerable water is pushed down the North Sea leading to very rough sea conditions. This situation accompanied by high tides can the lead to flooding as we had in 1953.

Occasionally I am asked where is the best place for good weather in the region. If we define good weather as reasonably dry with a fair amount of sunshine and being reasonably warm, I would say somewhere like Scunthorpe.

This is just a broad-brush description of our regional weather and of course in our daily weather there are many exceptions.

2 Extreme Weather!

Paul Hudson

The East Coast floods of 1953

The East Coast has always suffered from flooding. The Lincolnshire and Norfolk coastlines are particularly prone, being very flat and often reclaimed marshland. History is littered with stories of shipwrecks along this coastline, mainly in winter when the North Sea can become one of the roughest and most dangerous in the world. Natural or man-made sea defences protect the flat coastline, but on the night of Saturday, January 31st 1953, they were shown to be hopelessly inadequate to the onslaught of the biggest tidal surge ever recorded in the North Sea.

On Thursday morning, January 29th 1953, a developing area of low pressure just to the north of the Azores was picked up by Weather Explorer, one of the many weather ships owned by the Meteorological Office. This was routinely analysed by the observers and forecasters of the day – depressions of this nature are quite common in winter. By Friday morning it had deepened considerably, and the Met Office forecasts for the coming weekend predicted that this deep area of low pressure would hit Northern Ireland and western Scotland during late Friday into Saturday. It would then push southwards through the North Sea during Saturday night into Sunday, and gale warnings were issued around the coast. This sequence of weather events should not have caused undue alarm. But unfortunately by a cruel twist of fate, high spring tides were also forecast, and the resultant tidal surge caused one of the worst peacetime natural disasters the country has seen.

At about 7.45am on Saturday morning, Storm Force winds off Stranraer were already battering the coast, and as the Princess Victoria ferry struggled in her journey to Larne she met the full force of the gale's fury. Her car loading doors were forced open, and she began to take in seawater at an alarming rate. By around 2pm, having drifted for hours, she sank. The death toll was 132, which tragically included all women and children on board who had thought that they had escaped, only for the lifeboat to overturn as it was launched from the ferry. The unfolding catastrophe was to dominate the BBC news headlines for the rest of the weekend, but even as news began to filter down the coast, no-one had any idea of what was about to happen on that cold dark night.

That afternoon, ferocious winds were battering northern and eastern Scotland, with winds down the north-east of England beginning to increase. At the Met office at RAF Kinloss, a then record gust of 113mph was recorded. This was swiftly beaten by a gust of 125mph in the Orkneys, during a period of wind when the average speed was an

The North Sea can be one of the roughest and most dangerous in the world – and history is littered with stories of shipwrecks. This early photograph evocatively captures the sea conditions following the wreck off Whitby of the sailing ship Mary and Agnes in October 1885.

incredible 90mph – Hurricane Force 12. Further south along coast of Yorkshire, Lincolnshire and North Norfolk, as the sea increased ahead of the expected high spring tide, eye witness reports spoke of the fact that the previous morning tide had not receded and the estuaries abnormally full for low tide. But by teatime, the tidal surge had already torn through Scarborough's golden mile, lifting cars across the road as the sea rose to remarkable levels, even higher than the top of the harbour wall. Cleethorpes was next in line, the flat south bank of the Humber no match for the wall of water pushing ever southwards, as shops and properties were wrecked with winds speeds reaching 90mph. But apart from one or two fatalities, this part of the coast got away lightly.

That evening, the coast of east Lincolnshire saw a catastrophic failure of its sea defences, with Mablethorpe and Sutton-on-Sea bearing the full brunt of the sea. One can only imagine how frightening the situation was. It was a bitterly cold evening, with a Storm Force northerly wind and snow showers. Many people did not own televisions or telephones in those days, so although local people knew of the tragedy on the Princess Victoria ferry, they were mostly oblivious to what was happening. That evening, in the High Street in Sutton-on-Sea, water had been pouring through small gaps in the sea defence for half an hour, with ankle-deep seawater by that time. Within minutes, the sea defence collapsed and the street became inundated with three feet of water. The ice cold water (estimated temperature of 6°C) stopped at nothing, flooding houses and schools, vehicles and farms, as an air of panic set in among the residents who were stranded without electricity or heating.

One farmer, who lived alone on his farm 1½ miles inland, could have been one of the lucky ones. A friend knocked on his remote farmhouse door to urge him to evacuate because he had heard that a great wall of seawater was approaching. The farmer who had lived in the area all his life laughed and reminded his friend that his house

Uprooted lampposts and the remnants of chalets on Scarborough's Marine Drive testify to the destructive power of an East Coast storm. It happened on March 20th, 1964.

was 1½ miles from the sea, and there was no possibility of his home being flooded. By the time the tide had peaked, the sea had moved more than three miles inland. Residents recall their horror at the scenes that night. One lady, in trying to escape the waters with her baby, tragically lost hold of her child in the swirling sea. Others, understandably, were trapped upstairs in their homes with an air of panic, not knowing how high the sea would come. A fireman, one of the first on the scene that night, told of his utter despair at the magnitude of the rescue job that faced him and his only fire engine. In the end, 41 people, mainly in Mablethorpe and Sutton-on-Sea, perished that night.

Even worse affected was the coastline of North Norfolk, where the tidal surge reached a peak value of three metres above normal, the flat coastline's defences standing no chance against the wall of water heading ever southwards. One man, who until now had never told his story, said that he still to this day feels guilty having failed to save both his mother and father from the terrible flood. By the time the sea had returned to normal, over 300 people had perished in the flood in this country, with another 2,000 reported dead in Holland and Belgium.

This disaster was due to an amazing coincidence of events. Firstly, the Hurricane Force northerly wind would in itself cause a very rough sea. A north or north-west wind is the worst direction for the East Coast. Due to the force caused by the earth's rotation (known as the Coriolis force), the sea whipped up by this wind is deflected to the right, and so is pushed onto the coast. Secondly, the intense area of low pressure that moved southwards in the North Sea caused the surface of the water to rise, in the same way that an area of high pressure can force the sea surface down. And thirdly, there were the exceptionally high spring tides expected that evening – the North Sea tide always moves southwards eventually meeting the tide pushing eastwards up the English Channel.
The government in the wake of these terrible

The never-to-be-forgotten night of January 31st 1953, when a combination of high spring tides, intense low pressure and northerly winds touching Hurricane Force caused a natural disaster of appalling consequences.

(Top, and opposite page) *The trail of devastation left at Spurn Point by the wall of water driven south by the gale.*

(Lower) *A catastrophic failure of sea defences on the Lincolnshire coast meant that communities such as Mablethorpe bore the full brunt of the storm, with ice-cold water quickly flooding all in its path.*

floods set up the Waverley Committee, in order to prevent such a thing happening again. The report noted that, although 'strong winds of exceptional severity' had been forecast by the Met Office and that 'exceptionally high spring tides' had been forecast by the Hydrographic department, crucially neither thought to collaborate regarding their information. This would have hugely reduced loss of life if not of property and farmland. One of the main conclusions of the report was that both should work together, and today the Met Office headquarters has a storm tide forecast section whose function is to forecast such events in the future.

After the floods, coastal defences were rebuilt to a level that should withstand similar tidal surges. But the question is, could such a flood happen again? Looking back through history, a series of disasters has devastated the coast. One of the greatest on record was the so called 'St Elizabeth's Flood' of November 18th 1421, which cost 10,000 lives on the Dutch coast as over 70 villages were submerged. In 1607, the sea broke through in Norfolk, ruining farmland for years. On November 29th 1897, northerly gales again coinciding with a high spring tide caused serious coastal flooding, but crucially the peak surge was in the middle of the day, and so serious loss of life was avoided.

In years to come, we will have to cope with the additional problem of climate change. As the earth warms, our polar ice caps will continue to melt, and sea levels are forecast to rise alarmingly over the coming decades due to this melting and also due to the thermal expansion of the oceans. With or without climate change, it is only a matter of time before the next tidal surge affects the East Coast. However, the Environment Agency is now investing in improved coastal defences. Coupled with the much improved storm tide forecast facility at the Met Office and the far superior communications compared to 1953, this means that a loss of life such as we saw back on that tragic January day is highly unlikely.

The Met Office established a Storm Tide warning service after the floods of January 1953. Hull would no doubt be warned of the tidal surge of September 29th 1969, which led to these city gents struggling to work through flooded streets.

The violent wind

Our region is often buffeted by strong winds. The strongest are usually associated with active areas of low pressure sweeping in from the Atlantic, bringing heavy rain and gale force west or south-west winds. Occasionally I hear reports of tornadoes but thankfully, although there have been several occasions when parts of the country have experienced hurricane strength winds, this phenomena is very rare due to the fact that we live outside the normal latitudes of such storms. However, on February 16th 1962 much of Yorkshire experienced near 100mph winds that caused massive damage, during a mountain wave storm.

The Sheffield Storm

In the early hours of February 16th, much of the country felt the force of an Atlantic depression, with mean speeds in Scotland approaching Violent Storm Force 11 on the Beaufort scale. Across the North of England, many towns and cities such as Rotherham and Barnsley reported unexceptional mean speeds of 45mph. But shortly before dawn, the wind for no apparent reason increased to a mean speed of 70mph, with massive gusts as high as 96mph. These high wind speeds came at just the wrong time – the workers of South Yorkshire were either setting of to work, of perhaps finishing their night shift and returning home, when all hell let lose.

Chimney pots came crashing down, killing three people. A 40 metre high crane was thrown to the ground, destroying a nearby college. It was estimated that 100,000 homes were damaged in total, 100 homes were demolished due to the severity of the damage inflicted on them, and in one street a whole row of houses were literally blown apart. In the surrounding area 100 of the 250 schools were badly damaged and temporarily closed. The city of Sheffield was declared a disaster area, and a relief fund was set up and raised £35,000 for residents, some of whom had not taken out insurance for their stricken homes.
But it was not just Sheffield that suffered the terrible storm. In total eight people died and hundreds were injured across our region. Scores of people in West Yorkshire were simply blown off their feet, and Leeds City station was closed. But the countryside was affected badly as well. It was reported that 66,000 trees were blown over in and around the reservoirs of the West Riding and 160 acres of woodland were flattened so badly that it was decided that all trees were to be replanted. The grounds of Harewood House were badly hit, and Harlow Carr Gardens in Harrogate were left in a terrible state. Perhaps worst affected of all were the beautiful gardens of Chatsworth House, deep in the Derbyshire Peak District. Virtually all of the trees on this estate were uprooted by the storm, leaving an unbelievable trail of destruction. In fact the groundsman very accurately observed this event, taking a daily weather log which to this day is safely kept in the possession of the Duke and Duchess of Devonshire. Meanwhile, along the coast, winds peaked at 90mph around Whitby, with many caravans simply torn apart of turned over like toys.

The storm was the leading story for days on TV and Radio, with Richard Dimbleby broadcasting a live Panorama programme from the steel city of Sheffield. The Queen also sent a message of condolence:'Please convey my heartfelt sympathy to all the people of Yorkshire who have suffered as a result of the recent gales. I am greatly reassured by the reports of the magnificent and untiring help which the voluntary organisations have given to the civic authorities in their joint efforts to alleviate the immediate and widespread distress.'

So what caused such a violent storm? It was due to a phenomenon called mountain or lee waves. Overnight a warm front had pushed across the area, leaving us in what is called a 'warm sector'. This is an relatively stable air mass, often with a temperature inversion aloft – this is where an upper layer of air is warmer than the air beneath it, which acts as a 'lid' on the atmosphere. Cold air

"I call it bracing!"

is heavier than warm air, so the air blowing in from the west or south-west has no buoyancy and simply follows the contours of the Pennines. As the air rises up the slope of the Derwent moors, it continues to rise until it hits the temperature lid or inversion, at which point it bounces back towards the ground. A series of standing or stationary waves produce peaks and troughs.

Unfortunately for Sheffield, the city centre was amazingly positioned directly under one such wave trough, where the air had reached maximum velocity before bouncing aloft once more. It is very common in such a set-up for areas to the east – the Vale of York for instance – to record no wind whatsoever, being positioned under a peak, and here a characteristic wave cloud (otherwise known as Altocumulus Lenticularis) can often be seen (see back cover photograph). This is an excellent forecasting tool, the cloud being invariably associated with variable and gusty winds.

This lee wave phenomenon is probably the most common explanation and cause of the strongest gusty winds experienced in our region. These can be well forecast in general, but are virtually impossible to pin point as in the unique case of Sheffield. Here, the precise wavelength of the standing wave, together with the thermal structure of the atmosphere and the precise details of the land profile, all determined where the highest winds were going to be. Try telling the viewing public that Sheffield will have 100mph gusts whereas five miles down the road there will be no wind at all and you'll be sure to be carted off by the men in white coats!

Tornadoes

These are another phenomenon that can cause localised destruction. Put very simply, they are similar to areas of low pressure but on a very much smaller scale. The science of tornadoes is still not very well understood, and forecasting precisely when and where they may form is virtually impossible in this country. But it is estimated that wind speeds can reach 250mph in the centre of a tornado, and have been known to be powerful enough to move a 90-ton railway engine 50 metres along a railway line.

In our region I usually receive reports of at least one tornado every year. The ingredients for such development require a wind veer from the ground upwards, instability within the atmosphere and usually flat ground. For this reason, many reports have been just ahead and on passage of an active cold front in summer. In this situation, sharp wind veers of 180 degrees can take place, which, together with explosive uplift of hot afternoon air, seems in fairly rare instances to allow tornado development. Funnel clouds are the first stage of such development, and only on contact with the ground do these funnel clouds mature to be full-blown tornadoes. The vast majority of tornadoes are seen across flat ground, for instance in North Lincolnshire, and out to sea (where they are called waterspouts). This is because over hills or rough terrain the flow of air is disrupted by friction, and so the funnel cloud cannot reach the ground. The back cover picture of the funnel cloud near Ilkley is one such example of a near-tornado.

Most reports of violent tornadoes come from the Great Plains of the USA, which can offer the almost perfect conditions for such a phenomena. When a cold front, mainly in summer, heads southwards with its very dry air from the north it often meets very hot and humid air tracking northwards from the Gulf of Mexico. Where these two meet, we have a perfect mix of massive uplift of moisture laden air, sharp veer or turn on the cold front itself, and the very flat land. So forecasting tornadoes in this part of the world is much more straightforward.

The future

All wind-generating systems have heat as an energy source, whether it is a hurricane in the Caribbean whose heat source is the sea, or a tornado whose main source is the heat of the land. Our climate has been warming since the nineteenth century, and the last few years have seen global mean temperatures breaking all known records (which go back thousands of years). It is a sobering thought that as the seas get warmer, the hurricanes get stronger, and generally a warmer planet will mean weather systems have more energy. This will inevitably leading to stronger winds, more intense areas of low pressure and ever more powerful tornadoes across all parts of the world.

The floods of autumn 2000

Rivers exists for only one reason - to carry water from the hills to the sea on courses that were mostly carved into the land over 10,000 years ago during the last Ice Age. The starting points or sources of many of the region's rivers range from the magnificent River Aire, which has its origins in Malham Beck and the awesome limestone shadow of Malham Cove, to the breathtaking headwaters of the River Wharfe at Cam Flats in Upper Wharfedale. These babbling brooks grow in depth and width as they meander slowly south-eastwards through beautiful countryside before slowly joining with other streams and rivers to exit into the North Sea via the River Humber. In fact approximately one seventh of Britain's rainfall empties into the North Sea through this vast river. On occasions, there is too much water for the river network to cope with, and so the rivers naturally overflow into what we call 'flood plains'. These are areas of flat land, sometimes occupying the whole of the valley bottom, where the water can flow until the excess amount falls back to normal levels.

Communities built their homes along these rivers over the centuries usually out of necessity. The rivers were a source of drinking water, a mode of transport and were needed as industry began to

develop during the Industrial Revolution. But more recently pressure on land has meant that they have also become sought-after locations for homes and offices – with disastrous consequences. The autumn of 2000 turned out to be such a disaster.

Floods are nothing new to our region. Barely a year goes by now without land being awash with water, caused by a high water table due to persistent rain, or perhaps in summer when the land is dry during a hot spell – only for the heavens to open, leading to rapid run off and temporary flooding. But this is what we call 'localised' flooding. Traditionally, serious widespread flooding has generally occurred in late winter or early spring. Snow that has accumulated during the mid-winter months over the hills begins to thaw as warmer air starts to move in, which, together with its rain-laden cloud, adds to the melting snow to cause a sudden and prolonged flow of water from the hills. There is an old saying in the Dales that goes something like, 'There's nothing that can make a river rise quicker than snow melt.'

In fact it is often forgotten that, although there were serious problems faced by the region due to the harsh and prolonged winter of 1947, arguably more misery was experienced when the thaw and heavy rains arrived at the end of March 1947. But in the last few years, serious widespread flooding has been caused by rainfall alone, with an unprecedented intensity and regularity.

Following the celebrations of the new millennium, the calm weather conditions on the morning of January 1st 2000 could hardly have been a more misleading sign for the weather that was to follow in the next few months. The first few months of the New Year were largely unexceptional, with little snowfall and near average rainfall. However the first signs of trouble ahead came during springtime, with frequent spells of heavy rain during April causing waterlogged land and some localised flooding. In

Floods have long been a feature of life in the low-lying parts of Yorkshire.

(Opposite) *In June 1932 the River Don was some two miles wide and had almost engulfed the main East Coast railway line. The Flying Scotsman tiptoes past newly repaired track near Arksey.*

(This page) *Flooding of Ouse Bridge Inn on King's Staith, York, is a regular occurrence. A passing boat greets the first-floor occupants in February 1966.*

Even though the harsh winter of 1947 was exceptionally severe, the heavy rains that arrived at the end of March caused even more misery.

(Top) *The outskirts of Selby resembled one giant lake on March 26th 1947.*

(Lower) *Reminiscent of Venice – if not quite as beautiful! Houses in the centre of Selby receive a delivery of coal from a former landing craft.*

(Opposite) *York on March 24th 1947, seen from the air close to the confluence of the Foss with the Ouse. Floodwaters are lapping the approaches to Skeldergate Bridge.*

fact April turned out to be the wettest since 1756. The traditionally dry month of May failed to live up to expectations, with more rainfall - the wettest since 1983.

By the start of June, the land was on a knife-edge – Honley show, near Huddersfield, usually held on the first Saturday of June, was an early casualty of waterlogged land. At this time of the year there should be a net loss of water from the land, due to much increased evaporation by sunlight and due to moisture loss to vegetation. But further heavy rainfall during June finally tipped the balance, with many of our rivers placed on flood alert. The River Ouse climbed to its highest ever June levels causing more misery to the residents of York. Todmorden and Hebden Bridge bore the full brunt of a raging River Calder, as it bursts its banks and tore through homes with its filthy sewage and sludge. The clean up operation took months. But much, much worse was to follow.

The rest of the summer was unexceptional, with short-lived spells of warmth and near average rainfall. Crucially, the land did not experience any prolonged dry weather in June, July and August and as we entered the traditionally wet season of autumn, the water table was abnormally high. We desperately needed some dry weather. But what was to follow re-wrote the climate history books and set alarm bells ringing to the highest levels of government.

October had been unsettled, with spells of rain – it had been a fairly typical month until the last week, when rainfall began to turn more frequent and heavier. On Saturday October 28th, Wetherby racecourse was quickly turning into a quagmire as heavy rain and strong winds buffeted the area. On the afternoon of the October 29th torrential rain fell during the afternoon and evening – the televised football match between Bradford City and Leeds United graphically showing the near tropical intensity of the rainfall over Valley Parade as the pitch turned into a lake.

And then came the straw that broke the camel's back. On the morning on Monday October 30th a deep area of low pressure tracked right across our region – its central pressure just 958mb – producing another two inches of rain across the Dales in just a short space of time. Eye witness reports in Nidderdale reported thunder and lightning and blizzard like conditions, with temporary heavy snowfall reported across many parts of Pennine Yorkshire – all a direct consequence of the explosive development of the low and the huge fall in pressure associated with it. As the area of low pressure tracked into the North Sea a gust of 99mph was recorded on the Humber Bridge, with Hurricane Force 12 winds reported out at sea. By this time, understandably, the region's river network was simply unable to cope. By the end of the day virtually all the rivers were subject to flood alert, with 'severe' flood warnings being issued with alarming frequency. BBC Look North that evening reported scores of houses evacuated or flooded and thousands of animals stranded, in addition to the structural damage the storm had caused.

Further torrential rainfall fell during the cold early hours of Halloween 2000 as an intense convergence line developed from Keighley northwards to Malham. By the time this line of rain had run its course, thousands of people were flooded out of their businesses and homes. The list of misery was endless. Stockbridge in Keighley won the dubious accolade of having the highest concentration of flooded houses anywhere in Britain – 325. Houses along the River Don in South Yorkshire were ruined. The River Ouse in York

"I'm just going to get the milk!"

In recent years widespread flooding has occurred with unprecedented intensity and regularity. The record floods of autumn 2000 were in part due to an exceptionally wet September. It was nevertheless not as wet as September1981, which similarly had a knock-on effect at the beginning of 1982.

(Top) *A boat forms the only means of transport in Fishergate, Boroughbridge, on January 4th 1982.*
(Lower) *Two days later, this farm near Selby was photographed in a sea of water.*

reached its highest levels since records began in 1642. The towns of Gowdal and Barlby had never seen flood levels like it. And the towns of Malton, Norton and Stamford Bridge, which had only just recovered from devastating floods in March 1999, saw record floods again. The first two weeks of November saw no let up in this incredible sequence of events. At one stage the A1 northbound near Dishforth was under four feet of water, and some homes suffered repeat flooding as the river levels rose and fell sharply with every new band of rain.

In the end, September was the wettest since 1981, October the wettest since 1903, and November the wettest since 1970. Autumn 2000 turned out to be the wettest since 1872, and more rain fell during September, October and November than in any other three-month period since rainfall records began in 1727. By the time dry weather returned towards Christmas that year, millions of pounds worth of damage had been caused and families would be in temporary accommodation for months to come.

It is natural to want to find answers when such devastation happens. Why did we experience such floods? Who was to blame? True, drainage channels (commonly known as grips) at the top of the Dales do add to the flashy nature of some of the rivers as rainwater is encouraged to run off the land as quickly as possible. Blocking the drainage channels with peat bungs, thereby storing water upstream, can rectify this problem. Perhaps if some of our rivers were dredged then the rivers would have more capacity. But the truth of the events of 2000 is that the river network simply could not cope with such a huge volume of water. No matter how many 'grips' are blocked or rivers dredged, the rivers would still have burst their banks and spilled into the flood plains – as they are naturally designed to do.

Drought

Most weather extremes generally cause misery to those people unlucky enough to be afflicted by them. The East Coast floods claimed over 300 people across the country, hundreds of millions of pounds worth of damage was caused by the floods of autumn 2000, and 49 people died during the winter of 1962/63. But despite the problems that droughts invariably bring, the vast majority of the people often greet them with great glee due to the fact that they go hand in hand with memorable summers – and only in rare cases in this country do they lead to death or financial loss.

A notable exception to this was in 1666. The summer of that year was very dry, and by the end of August many streams and rivers were down to a trickle. On September 2nd 1666 a strong easterly wind developed over London. With all wood buildings tinder dry, a small fire in Pudding Lane quickly took hold with the resultant Great fire of London, now part of British history. By the time the fire was put out, 13,000 houses, churches and public buildings had been destroyed.

This region has been affected by three main droughts in the last century – 1959, 1976 and more recently 1995.

The drought of 1959

This was a beautiful summer, typified by unbroken sunshine for much of Yorkshire and Lincolnshire. Spring was very dry with persistent areas of high pressure over much of the country. Such high pressures or anticyclones yield little or no rain, but in this country it is quite rare to have long periods of such settled weather due to the fact that our location in north-west Europe exposes us to persistent areas of low pressure generated by the ever-present jet stream.

The drought of 1976 caused many reservoirs to become almost empty. Thruscross Reservoir in the upper Washburn valley was so low that acres of cracked earth were revealed, as well as remnants of the submerged West End village – including its packhorse bridge.

But by the August Bank holiday weekend, water stocks were dangerously low and rationing of water became a reality as standpipes were erected in the streets of our region. Many reservoirs were almost empty, with Silsden Reservoir near Keighley one of the first virtually to dry up – threatening the closure of the local dye works. Drastic action was taken, and water was eventually pumped in from the nearby River Aire, itself flowing at a dangerously low level. As is characteristic with such droughts, moorland fires were an increasing problem, and with many springs running dry, fighting such fires with water became almost impossible. The rains eventually arrived in autumn, which came as a huge relief to the authorities and by this time to the general public who became fed up with having to walk to the end of the road several times a day to fill up a bucket of water.

The drought of 1976

This is the year that most people refer to when the topic of long, hot summers is discussed. It is also the event that triggered the first tentative debate on global warming, a subject which seems to so dominate the headlines at the moment.

It is easy to forget that the seeds of the great drought of 1976 were sown back in 1975. This was also a very good summer, with below average rainfall and above average sunshine. Although September 1975 was wet, with above average rainfall, remarkably the rest of the year was much drier than normal, a trend which continued into 1976 – every month having below average rainfall with April, June, July and August 1976 exceptionally dry. The organisers of the Temple Newsam dog show in Leeds must have felt particularly aggrieved when their show was washed out in May – it turned out to be the only spell of heavy prolonged rainfall in months! Newspapers ran headlines such as 'Worst drought since 1720' and 'Today is hotter than Spain'.

By June, the water authorities became increasingly concerned as week after week of little or no rainfall began to take its toll on dwindling water stocks. Cereal crops in Lincolnshire and South Yorkshire were beginning to wilt under the relentless hot sunshine, with temperatures regularly over 90 degrees. Parts of the M1 began to disintegrate under the heat stress, and workers in the region's factories were sent home, unable to

(Opposite) *The River Washburn is reduced to a trickle as it flows into Lindley Wood Reservoir.*
(This Page) *Although at first a novelty, collecting water from standpipes soon became a chore. Mrs J Garside ('helped' by her daughter Joanna) and Mrs M Noble fill up their buckets in Primley Park Road, Alwoodley, on August 17th 1976.*

"They've run out of raspberry ripples!"

cope with the afternoon heat. There was even a story in the Bradford Telegraph and Argus from the Crown Court that Judge Gilbert Hartley gave permission for members of the bar to remove their wigs!

By the first week in July, hosepipe bans were in force. On July 12th a thunderstorm brought torrential rain to Bradford, but unfortunately the storm was very localised and missed most of the reservoir catchments. August got even hotter and drier, with clear blue skies for much of the month. Cereal and vegetable crops died due to lack of water, causing the price of food to rise and the county was almost on emergency footing as August ended.

Bob remembers the summer of 1976 very well. At the time he was a forecaster at RAF Bawtry and he was contacted on the Friday before the August Bank Holiday weekend by the Doncaster Evening Post. They said that they had interviewed a Greek lady who had been in touch with a Greek god, who had told her that the hot weather would come to an end over the weekend. In addition, there were to be thunderstorms and heavy rain that would bring an end to the drought conditions. Now at the time the Met Office forecast was for no change to the hot and sunny weather and so Bob politely suggested that the lady was talking a load of nonsense. What a disaster! On the Bank Holiday Monday thunderstorms developed widely across the region and there were reports of flooding. For the record, the Greek lady then made further predictions over the coming weeks, all of which proved to be wrong, which rather let the Met Office off the hook!

The great drought did in fact come to an end with a bang, with torrential rain into September and frequent thunderstorms. It eventually turned out to be the wettest September since 1727, and serious flooding was reported in many areas.

The summer of 1995

This drought was very different to those described above, and effectively strengthened the hand of the scientists arguing that climate change is real and will lead to big extremes in our weather. Unlike a classic drought situation, which usually requires the previous winter and summer to be drier than average, 1995 began with heavy rain and snow in January and February. In fact eighteen inches of snow fell over the hills at the end of January, when much of the region became gridlocked as snowfall that had been predicted came earlier than expected. The resultant thaw and continued heavy rain in February brought widespread flooding to the region, and, more importantly, groundwater levels and reservoir stocks as we entered March were high.

March and April saw the onset of much drier than average weather, the trend continuing into summer with August turning out to be exceptionally dry – with less than ten per cent of the normal expected rainfall. I was a forecaster at Leeds Weather Centre at the time, and the sequence of events was remarkable. Several cold fronts pushed across the region from the north-west, but pressure was so high that they gave no more than a band of cloud with a few spots of drizzle. As the fronts cleared the region, pressure rose and the area of high pressure became re-established.

The drought of 1995 was very much more localised than 1976. In Nidderdale, whose three reservoirs provide drinking water to Bradford, the dry weather conditions have been calculated to be a one in 200-year event if one looks at the period of May to September. But what was remarkable about this particular year was that the normal wet weather in September and October 1995 failed to materialise and reservoir levels remained dangerously low. When this is taken into consideration, the period as a whole is thought to be a 1 in 500-year event.

Coupled with the much-increased demand for water compared to 1976 and leakage from the underground network of water pipes, the situation became very serious indeed. Some reservoirs were within fourteen days of running out of water completely, and water was transported via the river network from Kielder Reservoir in the north-east of England to Yorkshire. In a report written by the Environment Agency to the Secretary of State on the then ongoing drought situation, it warned that if substantial rainfall did not fall during winter 1995/96 then the situation for the summer of 1996 would be even more severe. But once again the old weather saying proved to be right, 'Whether it be fine or whether it be wet the weather will almost always pay its debt' – as in the end the rains did return and water stocks eventually recovered.

Harsh winters

Our climate has always been extremely changeable. In the seventeenth and eighteenth centuries, winters were so cold that rivers such as the Thames regularly froze over. But it also has to be remembered that the Earth has actually undergone warmer periods as well. The Vikings colonised Greenland in such a warm period as the ice retreated northwards, and history tells us that vineyards were plenteous along the limestone crag of Knaresborough hundreds of years ago. As late as the 1960s and 1970s, some of the winters were so cold, and summers so poor, that scientists were beginning to wonder if we were beginning to slip into a new Ice Age.

But for anyone born in the last twenty years, it is perhaps impossible to remember how winters used to be in Yorkshire and Lincolnshire. Two of the commonest remarks that I hear are, 'We don't get winters like we used to', and 'The summers were much better when I was a lad!' The first of these remarks is certainly true but the second almost certainly is not. Whether the climate is going through a cyclical warm period, or, as seems much more likely, we are experiencing global warming, it is now common for winters to be snow-free and it is often rare to see regular frosts until well into the New Year.

The winter of 1962/63 was exceptionally harsh, with cold easterly winds and heavy snowfalls that cut off many Pennine communities. Workmen cut through snow on the Reeth to Brough road near England's highest inn at Tan Hill on January 11th 1963.

Harsh, cold winters are without exception caused by bitter easterly winds. It is worth remembering that our region is on the same latitude as Canada, but what stops us suffering such continental winters is the Gulf Stream which surrounds our shores with warm waters all the way from the sub tropics. Sometimes our weather pattern becomes blocked, and if the blocking area of high pressure is situated to the north of us, then we can pull in bitter continental easterly winds, sometimes with their origin in Siberia. Such anticyclones, once developed, can be very reluctant to move, as the air gets colder and denser with time and can affect our weather for much of the winter. Each winter described below was due to such atmospheric conditions.

The winter of 1978/79

The last real winter was in 1978/79, when trans-Pennine routes were regularly blocked, and temperatures were so cold that long-distance lorry drivers dangerously lit fires underneath their diesel tanks in order to defrost their fuel lines. January 1979 turned out to be one of the coldest this region has seen. On January 16th a state of emergency was declared in South Yorkshire, as the government minister David Howell flew by helicopter to Sheffield to see for himself some of the worst chaos anywhere in the country.

But it was the snowfall of February that ultimately grabbed the headlines across our region. Heavy snow and gale force winds, which pounded the region around the middle of the month, caused it to become almost cut off. The police set up 'Operation Siberia' to help out the gridlocked region, but even when troops came to their assistance, the elements defeated them, as even the main motorways were closed. Snowdrifts 30ft deep were reported along the A1 in North Yorkshire, and the York to Beverley road was blocked, as were all roads across the Wolds. Travel between Yorkshire and Lancashire, whether by road or rail, was impossible. On February 22nd, troops battled through to cut-off villages, and the now famous Hannah Hauxwell commented that she had never experienced anything like it.

The winter of 1962/63

This was described as the 'Long Winter' in the

A major rescue operation had to be mounted when the Edinburgh to London sleeping car express became snowbound near Dent in January 1963.

Guardian and was in general much worse than that of 1978/79. December 1962 had one or two cold snaps, but the winter famously began on Boxing Day. Cold easterly winds had already developed across the area, when an area of low pressure tracked across southern Britain, and with areas to the north of this in the cold air, copious amounts of snow fell countrywide. But it was January that proved to be a miserable month for many. Cold easterly winds continued to feed in from the continent, as a succession of low-pressure areas tracked to the south. These are the main mechanism for heavy snow, as moisture-laden clouds from the Atlantic meet the icy air from the east and turn what should be heavy rain into heavy snow.

As the easterly winds continued to rage, drifts became tens of feet high and communities were cut off. At one stage, firemen had to take drinking water to communities in Nidderdale, as village after village across the Pennines, North York Moors and Wolds became isolated. The coast was very badly affected, as winds gusting up to 70mph in Bridlington fed continuous snow showers in from the North Sea. By February many of the region's rivers, including the Ouse and remarkably the Humber, had simply frozen over. On the River Hull boats became trapped, and two seamen died in the Humber estuary due to the intense cold. By the end of February, Hardrow Force had become frozen for the first time since the end of the nineteenth century. By the time winter gave way to spring, 49 people across the country had died.

The winter of 1946/47

The start of 1947 was a cold one. The first snowfall came at the end of the first week, with Ilkley recording the coldest night in the country on January 7th, with an air temperature of -14°C (7°F)! A similar pattern of biting easterly winds and successive southerly-tracking depressions caused whiteouts, with hundreds of villages cut off from the outside world. Prisoners of war were mobilised across the area to help reconnect remote villages. The intense cold lasted through February, with some observing stations recording no sunshine for three weeks. By this time there

(This page) *The 1962/63 winter caused many waterfalls to freeze for the first time since the nineteenth century. The ice at Thornton Force, near Ingleton, was estimated to be five feet thick.*

(Opposite) *Ice skating, as seen here on Roundhay Park in Leeds, was a popular winter recreation in Victorian and Edwardian England when the colder winters led to thick ice forming on many lakes.*

was a sense of crisis within government as basic foodstuffs, already in short supply following years of war, became scarcer. The unrelenting cold and lack of heating caused hundreds of deaths. Much of the Dales was cut off by the end of February, and thousands of animals were buried alive under massive drifts.

By the last week of February, the Met Office stuck its neck out by forecasting milder weather. An Atlantic depression would track further north it said, with snow turning to rain followed by a general thaw. Wrong! What followed was described in the press as 'the Great Yorkshire Blizzard' - the depression had not got as far north as first thought. Nice to know I'm not the only one to get the forecast wrong!

The winter continued into March – all the region's rivers were by now frozen over to such a depth that ice-skating could safely be pursued. Relief finally came by mid-March, as milder south-westerly winds at last pushed across Yorkshire. But what followed – some of the worst floods in living memories as the snow melted – was in many ways harder to take for some people than the bitter winter that preceded it.

The question as to which winter - 1962/63 or 1946/47 - was the harshest is an argument that I hear regularly from people who experienced both. The mean temperature for both winter periods (December to February) shows that 1963 was the coldest (-0.1°C compared with +0.9°C) by 1°C. However, as is very often the case, it is the perception of the winter which is perhaps more important than the hard facts. Taking a closer look at the available statistics shows that although the two winters were similar from a snowfall point of view, there was one very big difference which makes people who still remember both winters adamant that 1946/47 was the worst - sunshine amounts.

Local and reliable accounts of both winters suggest that during the cold spell in 1947, which essentially takes in January, February and March, there were only six days when the sun shone, compared to 1963 when, in between the snowfalls, conditions were quite sunny. This in itself would account for the mean temperature in 1962/63 being colder due to the fact the nights would have been clearer, and hence frosts would have been more severe. In 1947, with skies continually overcast, the mean temperature would have been a little higher due to cloud cover at night, but this is misleading and overall the conditions would have felt colder. Psychologically it must have been a very depressing time with such an astonishing lack of sunshine, and for me it is clear that, contrary to what the statistics tell us, 1946/47 wins the accolade of worst winter of the century.

It may well be that winters as we used to know them are a thing of the past. As the Earth continues to warm, winters are forecast to be milder and wetter. But there may well be some hope for snow lovers, like myself, across the region. One independent theory suggests that our winters may well revert to a cold continental type after all. If the seas continue to become warmer, this will cause the ice at the poles to melt. This in turn will act to reduce the salinity of the Atlantic Ocean, which will cause the Gulf Stream to weaken. Some scientists predict that in time the Gulf Stream will stop short of our shores altogether, resulting in much harsher winters. We shall see!

(Top) *Horses used to come into their own in wintry conditions, as famously recorded in this funeral procession on the moors between Otley and Horsforth in 1947. Apart from heavy snowfalls, the early months of the year suffered from an astonishing lack of sunshine, thus earning 1947 the accolade of the worst winter of the twentieth century.*

(Lower) *Horses were also active in the earlier severe winter of 1940, in this instance conveying the milk by sledge at a Craven farm.*

3 Weather Forecasting

Bob Rust

How forecasting has developed

The first attempts at forecasting the weather were looking at the signs in nature. These were totally unreliable and something better had to be found. Most of the instruments for measuring the different elements of the weather had been invented and there had been many investigations into the wind patterns around the globe and the effects of the rotating earth on these wind patterns. Most of these investigations had been done between about 1650 and 1850 but there had been no real attempt at scientifically forecasting the weather.

The first such attempt was by Admiral Fitzroy (1805-1865), who was captain of the Beagle when it made its famous voyage around the world with Charles Darwin. Fitzroy was very keen to try and understand the weather and he spent his time making weather observations and studying the barometer. From his observations he noted that a certain type of weather usually followed falling or rising pressure. From these observations he came up with a few forecasting rules which became very popular and were often inscribed on barometers. These were:

A fall of half a tenth of an inch or more in half an hour is a sure sign of a storm.

A fall when the thermometer is low indicates snow or rain.

A fall with a rising thermometer indicates wind and rain from the southward.

Steady rise shows that fine weather may be expected and in winter there may be frost.

These are quite sensible rules and are a good guide to using a barometer at home.

Now apart from Fitzroy's work many places in Europe had set up weather observing networks. These included the Lunar Society of Birmingham, the Royal Society of London and the Societé de Medecine in France. The observers were encouraged to make simultaneous weather observations and keep weather diaries. By 1778 the Societas Meteorologica Palatine in Germany had fifty observers and this soon grew, with observers in Baghdad, New York, Stockholm and St Petersburg. At about the same time Thomas Jefferson, who later became President of America, and James Madison started to make simultaneous weather observations in Monticello and Williamsburg. In 1814 American army surgeons were ordered to keep weather diaries with a view to investigating the effect of weather on disease.

Now here with a basic observing network, we have the basis of a weather forecasting service. However there was one major problem – these weather observations could only be sent to one central area by mail. This could take a long time and any weather prediction that could be made from these observations would be way out of date. Weather forecasting relies very heavily on receiving weather observations very quickly.

Samuel Morse overcame this problem with the invention of the electric telegraph about 1850. Using the telegraph, centres were soon set up in many places to collect these observations and possibly try and make weather predictions. France and the Netherlands started their observation collection service in 1855 and the British service began in September 1860. Admiral Fitzroy became the Chief Meteorologist to the Board of Trade and soon afterwards many countries around the world followed suit. Over the next few years weather observations became standardised throughout the world and codes were devised to make transmission of the weather information easy.

In 1816, German physicist Henrich Brandes had started to plot weather information on a map and he produced the first weather map based on observations for the year 1783. The weather centres around the world used this technique of plotting observations on a map and all agreed on a standard format. The first published weather maps in this country were sold for one penny at the Great Exhibition of 1851, which was held at the Crystal Palace.

It was by using these weather maps and knowledge of physics and the atmosphere that the first forecasts were produced. Initially the fleet used them as a storm could cause havoc to ships, but in July 1861 the first forecasts for the public were issued in newspapers. These were based on twenty-two weather observations in the morning and ten reports in the afternoon from various parts of the country, plus a further five reports from the continent, but little information from the Atlantic, where most of our weather comes. With so little information the quality of these forecasts was poor but they were the best available and were welcomed by the public.

Sadly in 1865 Fitzroy committed suicide by cutting his throat. I don't know whether he did this because he got the forecast wrong, but at least we don't have to do this now if we are incorrect. Forecasts then ceased because the Royal Society thought that the current scientific knowledge was inadequate. The public were up in arms about this decision and forecasts were started again in 1876. These forecasts were relayed to the public by press, telegram and telephone.

Progress in forecasting over the next few years was quite slow, although more weather observing stations did open. One big problem was the lack of information from the upper atmosphere. Many attempts were made to obtain measurements using kites, balloons and even manned hot air balloons. It was the manned balloons that provided the first useful information about the upper atmosphere. As early as 1783 a French physicist went aloft with a barometer and a year later Dr John Jeffries and Jean Pierre Blanchard made a flight to 9,000 ft, making observations of temperature and pressure. Many more and higher flights were made, and in 1901 one flight reached an altitude of 35,435 ft. A Swiss physicist reaching a height of 53,153 ft broke this record in 1931.

All of these flights provided some understanding of the upper atmosphere but to be useful to forecasters they needed to be made much more often, which was impractical.

Some useful information could be obtained by taking observations up mountains and an observatory was built on the top of Ben Nevis. A couple of years before the observatory was built

Weather forecasting was transformed in the 1930s with the introduction of 'radio sondes' – lightweight transmitters that were hung from hydrogen-filled balloons and released into the atmosphere. Obtaining weather data in this way was a regular duty for the stationmaster at Ribblehead, an exposed location on the Settle to Carlisle railway and noted as one of the wettest and windiest places in Yorkshire.

some instruments were installed on the summit of the mountain. Between June 1st and October 14th 1881, a Clement Wragg, who later was in charge of the Australian Meteorological Office, made daily weather observations on the mountain. He started with an observation at ground level at about 5am. He then climbed the mountain making regular observations and reached the summit about 9am. Whilst on the top he made a few more observations. He would start to climb down at about 10am making more observations en route, and reached the bottom about 1pm. To help him with this rather arduous task he was supplied with a horse, which took him halfway up the mountain, and a trained assistant did the job twice each week. Such dedication!

During the First World War there was an increased demand for more accurate weather forecasts including use for artillery, aircraft and observation balloons. This did bring about some improvements and an expansion of the weather service. It was also during this time that a major step forward was made which is the basis of present day weather forecasting. Lewis Fry Richardson, who was an eminent physicist and meteorologist, served in the war, and during that time he gave much thought to the problem of weather forecasting. Now weather is not just a jumbled disorder – it follows all the laws of physics. Richardson worked out a method by which weather could be predicted using mathematical equations. He decided that if you could describe the initial conditions of the weather using two thousand regularly spaced weather stations over land and sea, you could then predict weather changes using these mathematical equations. His paper on this subject was published in 1922.

Unfortunately this was only to be a dream of the future as the solving of these equations was not practical in those days. It has been retrospectively calculated that it would have taken an army of 64,000 mathematicians working twenty-four hours a day for a whole year to forecast the surface and upper air weather for just one day ahead. The good thing however was that Richardson did live long enough to see that one day his ideas would be the basis of all future forecasting.

Throughout the twenties, thirties and forties more improvements developed with more observing stations coming on line. Many techniques were devised for forecasting such things as maximum and minimum temperatures, fog formation and clearance temperatures and hence the times at which this would happen. The problems of gathering data about the upper air were overcome in Britain with the introduction of radio sondes in 1937. Basically these were lightweight radio transmitters which contained meteorological instruments that could measure temperature, pressure and humidity and send this information back to a ground station as a radio signal.

These instruments were hung from a hydrogen-filled balloon and launched into the atmosphere. As the balloon rose through the air it would be blown by the winds. Knowing the rate at which the balloon would rise, you were able to calculate the wind speed and direction at different levels by tracking it. As the balloon roses it got bigger because the pressure decreases with height. Eventually the balloon would burst and the instruments would fall back to earth, along with the balloon remnants and a radar net used for tracking purposes. To slow down the fall they were equipped with a small parachute. Most of these were lost, often finishing up in the North Sea or beyond because of the strong upper winds. However occasionally a few were recovered and they could be returned to the Met Office, who gave a reward of about ten shillings (50 pence).

It was also during the 1940s that computers started to be utilised for weather predicting using Richardson's ideas. The first attempts were made at Princeton University in America using a computer known as MANIAC – perhaps a good name for some present day forecasters. It took ten years of work to get some acceptable results but this was to be the basis of all present day forecasting.

In October 1959 the first satellite for providing weather information was launched and then others followed at regular intervals. However it wasn't until February 1966 that ESSA 1 first transmitted cloud pictures back to Earth. I can remember that at Bawtry they didn't receive satellite pictures but they did get a drawing of the clouds which someone had done.

Prior to the regular use of computer-generated forecasts most of the forecasting was done by chart analysis. Plotted surface charts and upper air charts were beautifully drawn and analysed. Movements of systems were determined along with tried and tested techniques for detailing the weather. Experience and local knowledge then came into play, eventually producing weather forecasts.

Over the past thirty years or so the demand for specialised forecasts has vastly increased. Luckily this has coincided with giant steps forward in computer development and a wide variety of other technical advances.

The present day forecast office consists of a multitude of computer screens and very little paper. Computers are used to model the entire atmosphere around the globe and predict development up to ten or more days ahead. This really gives the broad scale developments around the globe. For more detail a lesser area is used, taking in the Atlantic and much of Western Europe and an even smaller area model provides most of the detail we receive in our daily media forecasts. With this model, predictions of all the elements of the weather are made at intervals of as little as five kilometres. In the future it is hoped to reduce this distance to one kilometre making for even more accurate forecasts. It is the vast amount of data to analyse and the many calculations to be done that determines the chosen area. As computers become faster and can handle more data, forecasts should continue to improve.

These computer products are really the forecaster's plan of what should happen. It is not assumed that these are perfect and progress is continuously monitored. By keeping a watchful eye on the actual weather, using observations, satellite pictures and rainfall radar, the forecaster is able to check that everything is running to plan. Some adjustments may be made using the forecaster's experience and taking into account local effects. From all of this information, forecasts are prepared for the many and varied users.

TV weather forcasts

Over the years there have been considerable changes in television forecasts. The first forecast shown on television was on November 1st 1936 and transmitted by the BBC from Alexander Palace. This comprised just showing a hand drawn weather chart with the isobars and pressure centres. In those days this was just in black and white with a voice reading the forecast. There was very little change until the outbreak of the Second World War in 1939, when all weather forecasts ceased as they were classified as secret.

It was not until the June 29th 1949 that television forecasts were recommenced. Again these were just hand drawn weather maps with captions over them and a background voice reading the forecast.

We had to wait until 1954 to see a television forecast presented by a professional weather forecaster. He was George Cowling and worked for the Meteorological Office. It was decided that he should receive some extra payment for appearing on television, and he was paid ten shillings (50p) for each appearance. Other television presenters I recall from the early years are Jack Armstrong, George Luce and Geoffrey Leaf before we came to the era of Bert Ford, Jack Scott and Graham Parker. Again there had been little progress with the graphics and forecasters still used hand drawn charts. One of the first problems that the forecaster encountered was what we call 'Met Man's back'. This was the problem that the forecasters turned to look at the charts and give a forecast and explanation, and finished up talking to the chart with their back to the viewers.

It wasn't too long before magnetic tape and plastic was introduced into the presentations. It was a major step forward, in that it was easier to produce smoother lines on weather charts and symbols were made which could be removed or added to the chart. Being able to move symbols was a big improvement in that it was possible to show how the weather was expected to change. This did however present a few problems. It could be difficult to lift off some of the symbols and when moved they didn't always stick on the chart – some fell off and others occasionally slid down the chart. This may not seem like a major problem but to the forecaster it was a nightmare. Just think of the poor chap! He would be ad-libbing the forecast, trying to move his magnetic symbols and keeping an eye open for a man in the studio, who would be giving him signals of how much time he had left to finish the forecast. On the occasions when things do go wrong on a weather presentation, such as symbols falling off, it is not all bad news. Many forecasters have received a payment for these clips to be shown on other light entertainment programmes.

For many years the BBC national weather forecast was probably the Rolls Royce of television forecasts. Most of the maps still used the magnetic tape and symbols but the size of the maps increased and the now more professional forecasters would walk from chart to chart, avoiding the 'Met Man's back'. Satellite pictures were later introduced but the quality was often rather poor and they were only in black and white.

Other television companies were trying many other methods of weather presentation. One popular one was to use a revolving cube. This was fine in itself, but it did mean that you could only show four pictures to tell the weather story for the next twenty-four hours.

All of these forecasts were very popular with the viewers and there was an increase in the number shown each day. It was said that when the BBC had their evening news, the number of viewers used to rise towards the end of the news as more people tuned in just to see the weather forecast. There were usually only two or three forecasters on television and they became household names. In those days most of the men on television were soberly dressed in dark clothes, but a few of the weathermen started to wear somewhat brighter outfits. Initially the weather presenters were all male but eventually females were introduced and, if memory serves correctly, the first one was Barbara Edwards.

Nowadays we all have colour television and we are used to seeing weather fronts depicted in colour. Warm fronts are red with red half circles, cold fronts are shown in blue with blue triangles and occlusions are either shown as red and blue or purple with triangles and circles. The origin of this really dates back to the introduction of weather forecasts on television. Forecasters usually draw fronts on their working charts as just red, blue or purple lines. With only black and white available on television at first, it was decide to add the half circles and triangles to distinguish between the fronts. They were drawn on the front edge of the front and were therefore useful to show the direction of movement. This is why they have been retained even with colour television.

Initially only the BBC had presented forecasts but they then appeared on ITV's morning television, and later on some regional television, both independent and BBC.

In the early days all the television weather presenters were qualified meteorologists employed by the Met Office. The first forecaster on GMTV was Commander Philpott, who had been employed by the naval meteorological service. Tyne Tees Television was one of the first companies to have a regional weather forecast and they also broke the mould by having a non-meteorologist presenter, Wincey Willis.

Television producers then seemed to like the idea of glamorous weather forecasters and we started to see more young ladies on our screens. Some used this opportunity as a stepping stone to other roles on television, probably the most notable one being Ulrika Jonsson. More females appeared on the BBC weather forecasts but they were still trained forecasters. It is only over the past few years that they have used a few presenters who are not practising forecasters. However no names, no pack drill.

The next big change in television weather forecasts came with the introduction of computer-generated graphics. This was a big improvement, in that the forecasters could use many more charts to explain the forecast. All of the charts were prepared much more quickly and text could be easily added if necessary. Although we now had very professionally presented forecasts with excellent graphics, further development continued. With more powerful graphic computers available, animation was introduced. It meant that we could actually show rain falling from moving clouds in addition to animated satellite pictures and rainfall radar. Nowadays the Met Office forecasting computer is directly linked to the television graphics computers making it very easy to produce the charts. With the addition of animation we are able to see the changing wind patterns and different colours show temperature changes.

More and more information is given in weather forecasts with the inclusion of text warnings, flooding prospects, pollen counts and sun indices. Numerous forecasts are now available throughout the day on many television channels. The majority of the forecasts originate from the Meteorological Office, although there are a few that come from private weather consultants. This is why there are occasionally some differences between the forecasts.

With all this information available one has to ask the question, does the public really understand weather forecasts? This subject could be debated for hours and many investigations have been carried out into the public's reaction. From my experience on television, meeting viewers and receiving letters, the first problem seems to be that many people do not know where they live on a blank map. Many years ago one university carried out an investigation asking people in the street to point out where York was on a blank map of the British Isles. The results were quite staggering. Hardly anyone got it right and in a few cases they were as much as 150 miles out. This really is not surprising, in that unless there is a physical feature such as say the River Humber estuary it is difficult to have a starting point. Although it is rather extreme, I have received letters in the past asking me to show which is east and west.

Assuming that you do know where your town or city is, unless there is a weather symbol directly over the area, some people do not know what type of weather to expect. This problem has now been largely overcome with animated graphics, as most people do see the weather symbols moving over the area where they live. The main problem nowadays seems to be that we are supplied with too much information to absorb and understand in a rather short space of time. Many forecasts show animated weather for several days ahead and it is quite easy to miss out on the detail for the next 24 hours, particularly on national forecasts. Here regional forecasts do have an advantage in that they are looking at a smaller area and can provide more detail in the limited time available.

One technique often used in television weather presentations is chromo key. With this, it appears as though there is a weather map behind the presenter, although in truth there is just a blank background. So although the viewer can see the map, the forecaster has nothing behind him to look at. This is overcome by having a television monitor to the side of the weather set. Although the forecaster is really pointing to a blank board, he sees both his hand and the map on the monitor. If you look very carefully at many weather presentations, you should see that the forecaster's eye line is just off the map as he looks at the television monitor. One problem with this method is that if the background colour is say yellow, the forecaster cannot wear yellow as he would vanish from the picture.

As we all know, forecasts do occasionally go wrong and no meteorologist will claim infallibility. For a twenty-four hour forecast the Meteorological Office claims an accuracy of about 80 per cent. However when considering accuracy, it does look at several elements of the forecast, such as precipitation, temperature and wind. This method is correct, as many customers are perhaps only interested in say temperature or wind. So it is possible to score a high accuracy figure on a dry day when rain has been forecast. However to the man in the street, who may only interested in whether it will be wet or dry, the forecast is incorrect.

One method of predicting the weather is persistence – namely that is what is happening today will happen tomorrow – and this should give an accuracy of above 50 per cent. In 1987 some students from Sheffield Polytechnic decided to test this theory on rain alone. At the same time Leeds Weather Centre did the same trial using Meteorological Office methods. The results were rather surprising. The Leeds Weather Centre correctly forecast the rain in Sheffield on 67 per cent of occasions whereas the Polytechnic figure was 65.7 per cent. Although this was just one test it does show that using persistence will work on more than half of occasions.

If you have a little understanding of weather maps you can interpret them yourself. To begin with they show the centre of pressure usually marked 'High' or 'Low'. With high pressure it usually means settled weather and low pressure is mostly associated with unsettled weather. Surrounding the pressure centres we have the isobars. An isobar is a line on a weather map that joins all the places having the same pressure. These are very much like the contour lines we see on maps, showing the height above sea level. Now the isobars are an indication of wind direction, in that the wind blows virtually along the isobar, with low pressure to the left. Put more simply, the winds blow anti-clockwise around low pressures and clockwise around highs. In the southern hemisphere this direction is reversed. How closely spaced are the isobars gives a measure of the wind speed. If the isobars are tightly packed the wind will be strong, whereas widely spaced isobars mean light winds. The direction of the wind can give a good indication of expected weather. Between north-west clockwise through to north-east usually means cool or cold weather in our region. South-westerly winds bring mild but unsettled weather in both summer and winter and east to south-east winds usually mean very cold and dry weather in winter but fine and often warm conditions in summer, although in our region the temperatures are somewhat depressed by the North Sea.

Weather fronts, that is our blue, red and purple lines show where warm air is battling with cold air and they all indicate cloud and rain. A warm front usually means cloudy skies and a long period of occasionally heavy rain, which will be followed by rather cloudy conditions with perhaps a little drizzle and rather poor visibility over high ground. The blue cold front often gives a shorter period of rain, which may be heavy and sometimes accompanied by thunder. Behind the cold front clouds should break to give plenty of sunshine with an increasing risk of some showers later.

The occlusion, which is depicted as a purple or red and blue line, is where a cold front has caught up a warm front. This always happens because the cold air moves more quickly than the warm air. After the occlusion process the weather system slowly decays and dies. Occlusions usually bring us a band of cloud with some rain, which is often heavy in a newly forming occlusion. Occlusions often are moving slowly and, although they may not have a wide area of activity, they can bring a long period of cloud and rain. So by looking at the television weather maps, one is able to have a good idea what might happen to the weather.

The terminology weather presenters use to describe the weather can be confusing. Looking firstly at temperature, we often hear that 'it will be warmer or colder than average'. To most people this is meaningless unless they happen to carry round a table of average temperatures. The most sensible approach to this is to compare the expected temperature with that of recent days. So to say 'tomorrow will be a little warmer than today' will be more widely understood. With low temperatures and strong winds, most television forecasters make great play about the wind chill effect. A chart of these wind chill adjusted temperatures is often shown, but in my opinion these can confuse and it would suffice just to say, 'The wind will make it feel much colder.'

Many people are confused by the terms frost and ground frost. The explanation of this is that at night, the ground loses heat and cools and then this cooling is passed on to the air. It usually means that minimum air temperatures are a little above those on the ground, so it is possible to have frost on the grass and yet the temperature of the air is above freezing. The opposite happens on a sunny day in that ground temperatures are higher than air temperatures. You can easily see this by feeling how warm paths and pavements are on a hot summer's day. It is from this effect that we say 'it was so hot, it was cracking the flags' or 'you could fry eggs on the pavement'.

Wind speeds expressed in mph or Beaufort force are over the heads of most people, although they are understood and used by sailors and fishermen.

The terms rain and showers occasionally cause confusion, as to the viewer they both wet you. Rain means a period of mostly continuous precipitation with little or no brightness, whereas showers are short periods of rain mostly lasting only a few minutes and then followed by sunny skies. Sometimes showers gradually turn into continuous rain and many forecasters say the rain will become more organised. I feel that this is a fairly apt description, although I have heard many people ask, 'What is organised rain?'

With continuing competition between television channels for viewers, we no doubt will see further advances in the way weather forecasts are presented, but let us hope that they do not deliver so much information that they confuse more than they inform.

Finally, consider that television forecasters give their presentation without any script and throughout the forecast they are being told through an earpiece how much longer they have to speak. Also, during all this they turn and touch a map that often isn't there and give the odd smile to the camera, so they really do a pretty good job.

Thirsk moles hear Bob Rust's forecast!

4 Bob Rust - TV Forecaster

How I became a weather forecaster

Growing up as a young boy in Doncaster, I did not have any thoughts of becoming a weather forecaster in the future. In fact for the first few years there were no weather forecasts. This was because the war was still going on and weather forecasts were secret. The government didn't want to let Mr. Hitler know that it would be a fine clear night for his bombers. After the war I would occasionally hear the shipping forecast on the radio which would talk about strange places with names such as German Bight , Cromarty and Heligoland but even this did not inspire me to want to be a weatherman when I grew up.

Of course we were very much aware of the changing weather, in that it would dictate what we would wear and whether we could play out or not. Looking back, it did seem that the winters were much colder than they are now and we could usually rely on some snowfall. Like for the youngsters of today, this used to be exciting and a lot of fun. One thing most of us did learn, was not to walk on some pavements when there had been snow. This was to avoid being hit by snow sliding from the roofs of the rows of Victorian terraced houses. It wasn't necessary to have a thaw to bring the snow crashing down. The houses then had little or no roof insulation and with roaring coal fires, the unfaltering heat loss through the roof soon got the snow on the move.

Throughout my early and mid-teens I still had no interest or knowledge of the weather. At school, some aspects of weather and climate were taught in geography. Now I attended Doncaster Grammar School but I dropped geography after two years and at that point they hadn't reached 'weather' in the syllabus. It was quite amusing many years later when I was teaching meteorology to some amateur sailors. One of the students was my former geography master. I was only sorry that I couldn't even things up and give him the cane.

After leaving school I got my first job as a laboratory technician with the local gas board. Looking for a little more money I then moved to the National Coal Board's laboratories. The work was enjoyable and interesting, involving testing all aspects of mine safety and coal quality.

I then heard that the Scientific Civil Service had a Safety in Mines Research Establishment at Sheffield and the salaries were higher than the Coal Board's, so I applied to enter their annual Open Competition. I was successful and started working at Sheffield. Although I was now approaching twenty, I still had no knowledge or interest in the weather. To me the Meteorological Office comprised the few chaps we used to see on television and hear on the radio giving weather forecasts.

It was whilst working at Sheffield that I met some young men who worked for the Meteorological Office as weather observers. I discovered that the Met Office was also part of the Scientific Civil Service and came under the Ministry of Defence, whilst my parent department was the Ministry of Power. I was also to learn that promotion in the Met Office was a little easier than in my department. One slight downside was that their job involved working shifts and weekends. This of course meant that if I applied for a job and was successful, I would not always be able to see my beloved Doncaster Rovers. However, when told that there was extra pay for shift working and the fact that Rovers weren't really that good, I decided to apply.

As we all worked for the same employer, the government, an interview for a transfer was soon arranged. I was asked to attend Bawtry Met Office for an interview. This was a rather short interview. The senior meteorological officer soon realised that I knew nothing about the workings of the weather and he knew nothing about mine safety so we didn't have a lot to talk about. Nowadays I would probably have been made the Chief Executive but back then I was just offered a post as a weather observer. I did find out later that the Met Office was desperate to recruit weather observers as many young people did not want to work shifts.

So this was the beginning of my career in meteorology. I was posted to RAF Lindholme as a weather observer. The job wasn't that exciting or demanding, looking at clouds and reporting whether it was raining or not. One interesting thing was that there was supposed to be a ghost at the airfield. His name was Billy Lindholme and it was said that he was a wartime pilot who had crashed into the peat bog adjoining the airfield. There were many tales of him being seen and even speaking but I never encountered him. You can imagine that when working nights and you were the only person on the dark airfield, any slight noise was quite disturbing and you couldn't wait for the sun to rise and other people turn up to work. A few years ago the body of a wartime airman was recovered from the peat bog and one wonders whether or not there was a Billy Lindholme.

During my time as a weather observer at Lindholme I did become very interested in the weather and after a couple of years I was promoted and sent to the Met. Office College to be trained as a forecaster. My first job as a forecaster was at RAF Waddington. This was a Bomber Command station, which flew the Vulcan bombers carrying our nuclear deterrent. The job was very interesting and involved preparing forecasts for both high and low level flying. One aspect of the job that I believe helped me later as a TV forecaster was the morning briefings. The met man' was required to stand up and brief the station commander and aircrew on the expected weather for the day. Unlike television, my audience then could ask searching questions about the forecast. This taught me to quickly think on my feet and at least sound like an expert. This reminds me of some advice I was given many years ago, and that was, 'Sound confident. You never know you might get it right.'

From Waddington I moved back to Lindholme for a time and then on to Bawtry Met Office, where I stayed until it closed down and then I moved on to Leeds Weather Centre.

Bawtry was responsible for all the weather forecasting requirements for most of north-east England. We not only prepared aviation forecasts, but also forecasts for the press, radio, agriculture, commerce, industry and the general public. In fact any activity that had a weather connection. On summer weekends we had to prepare a lot of forecasts for pigeon racing. On occasions these could be disastrous. If the birds did not reach home on race day, the weather forecast was usually blamed. I believe that I hold the rather undistinguished record of losing in excess of three thousand birds on one day. On that occasion the weather forecast wasn't at fault and a number of birds did arrive home a few days later. I think that many of the losses were down to the accurate shooting of the French sportsmen wanting pigeon pie on the menu.

My long career as the weather presenter on Calendar began with a taped audition with Richard Whiteley. On a later occasion, I am telling Richard about forthcoming weather prospects in an outside broadcast from the top of Emley Moor transmitter.

It was during my time at Bawtry that my broadcasting career began with local BBC radios, Leeds, Humberside, Sheffield and York starting to take live forecasts. This meant that you started to get your name known around the region. Around the mid-1970s, Yorkshire Television started trials for breakfast television and decided to have weather forecasts. We did not appear on TV but delivered the forecast by telephone. Our picture appeared on the screen with a captioned name, and whilst giving the forecast the picture would be changed to another weather-appropriate one. Mistakes did happen and it wasn't unusual for a picture of a field of contented cattle to be captioned as 'Bob Rust –weather forecaster'. Despite these little problems it did mean we got publicity and a chance to speak to such famous people as Richard Whiteley and Geoff Druett.

It was whilst at Bawtry that I did make my first television appearance and this was on the BBC. On a couple of occasions the BBC sent Khalid Aziz down to Bawtry to do a piece about the Christmas weather prospects. Being the duty forecaster I was the interviewee. The BBC also made a series called Home Town and Bawtry was one of the places included. They sent the cameras and a reporter, Ken Cooper, to the Met Office and I had to explain to him how we forecast the weather. It was quite exciting watching the programme when it was transmitted some weeks later.

About the time it was announced that Bawtry would close and be replaced by a new weather centre in Leeds, Yorkshire Television approached the Met Office to ask if they could supply people to present the weather on Calendar. Prior to this they had the occasional appearance of Bill Foggit, who, clutching a piece of Winter Jasmine and a barometer would explain that that the activity of moles and goldfish meant that we were in for a bad winter. The Bawtry forecasters were asked if they were interested in the job and most volunteered to attend an audition. The main reason for so many volunteers was not the attraction of appearing on television, but that if successful they were promised a posting to the new Leeds Weather Centre.

We were invited along to the YTV studios and had to do a weather presentation with Richard Whiteley. Although we were given a make-up to improve our appearance, we received no other training or advice about appearing on television. These auditions were taped and several weeks later I was told that I was one of five who had been accepted. The service started and after a year we were weeded down to just two of us. By the end of the following year I was the only one left. I did not suspect that I would be there for the next twelve years with about ten thousand forecasts behind me, which is a lot of television. Later in this chapter I will tell you how the presentation changed over the years and recount a few of the lighter moments.

It could have been that I would never have appeared on TV had circumstances been different. I did learn after I got the job that Yorkshire Television were really looking for a lady weather presenter at first. One young lady was auditioned but the then boss of Calendar decided that she wasn't really television material. Some of you may have heard of that young lady, her name was Carol Vorderman. I wonder what she went on to do.

My fourteen years of television

Having presented something in excess of ten thousand forecasts on television, I obviously don't remember all of them but a few are worth recalling. Understandably I still remember the first one. For the whole of the week before I made the first appearance, it was rather worrying, knowing I would be appearing live on a channel that boasted six million viewers. Not knowing what the forecast might be on the night I was unable to practice what I would say. In fact for a few nights I didn't sleep because of the worry. I did think that perhaps I should reconsider my decision to appear. On the Saturday before I was to appear on the Monday, my wife Shirley took me into Doncaster to buy a new shirt and tie, thus assuring I would look presentable. In those days most people who appeared on television were smartly dressed. Monday soon arrived and a taxi had been arranged to take me from Bawtry to the studios of Yorkshire Television at Leeds. So off I set, clutching a satellite picture and a hand drawn weather map.

I was warmly welcomed on my arrival at the studios and was first taken to make-up where I was given a good powdering and my hair combed. I actually had some hair when I started doing television but very little at the end of my career. I met the director and producer of Calendar and it was sensibly decided that, as this was the first night for me, and also the first night they had ventured into weather presentation, it would be best to have a rehearsal. It would seem that there hadn't been a great deal of thought put into how the weather would be presented. After some lengthy discussions it was decided that my two charts would be fastened to two sides of a caption stand, which is very much like a music stand.

Marylyn Webb then arrived in the studio as she had been selected to do the weather interview. All that I had to worry about was giving the forecast and she would control the timing. I must mention here that timing of all items in television is very important. There is a time slot for the programme, and that must be strictly adhered to, otherwise Calendar might run into Emmerdale. We had a practice run and the director said that it was very good so I felt pleased. Feeling happy with myself I was then shown into the Green Room. This is where all the presenters and guests gather before the programme starts to have tea and biscuits. I am told that in the past they did supply something stronger! In the Green Room there was Richard Whiteley, and I believe Alan Hardwick along with Marylyn Webb, so I felt that I was among the rich and famous. However the real bonus was that Spike Milligan was a guest on the programme. I never imagined that I would be sat next to famous people chatting over a cup of tea.

As the weather was at the end of the programme, I wasn't taken back into the studio until after the commercial break. I stood in my position and eventually I was joined by Marylyn Webb. It wasn't long before the camera lined up in front of us and at this point my heart was racing. Marylyn introduced me as the new weatherman and then asked me what the weather prospects were for the next twenty-four hours. Off I went into the forecast, pointing out the features on my weather map and satellite picture, and then talked at length about the prospects for that night and the next day. It seemed as though Marylyn was trying to interrupt me but I had not yet finished, so I kept talking until I got to the end of the forecast. What had happened was that I had overrun and the programme finished at a rather abrupt end. Despite this I was quite pleased with my performance. Looking back however, it was probably rather poor.

Arriving home I was on something of a high and my family told me that I had been very good and looked nice in my new clothes. A few telephone calls from relatives also confirmed this, so I felt very pleased with myself and was ready for the next appearance, but this wasn't going to be for about five days. Remember there were five forecasters at first. The following day I went shopping in Doncaster with my wife. It occurred to me that if there were six million possible viewers, then there was bound to be one or two in Doncaster and they would probably recognise me as being off the telly. However, no one did and it was quite a long time before I was recognised. This was quite enjoyable at first but later on, when it was happening all of the time, it did become something of a nuisance. I will tell you more about that later.

Over the next few months or so things continued to go quite well with the method of presentation changing, until we settled with the weather cube. Compared to presentations today, this was really in the dark ages. We displayed a large satellite picture on one face and a hand drawn weather map on another. The other two faces had regional maps, one for that day and one for the next. We would give an explanation of the weather development using the chart and satellite picture and then talk about the regional prospects using the other two maps. There were no symbols on the maps – just the names of a few towns and cities. The timing for the presentation was still controlled by one of the presenters and we both sat at a desk in front of the weather cube. When our time was nearly up, the presenter would say, 'Well, give us the summary now', and we would finish with a ten second summary. When things weren't going well and there was a shortage of time, you could be only part way through the forecast and suddenly you were asked, 'Give the summary.' Although such occasions were rare, if there was a lot of time and you had finished the forecast, the presenter had to earn his corn and ask a supplementary question such as, 'What is an occlusion?'

The first year was really just a settling in period with little or no chat or humour in the weather, although they did start to introduce a few forecasts from outside locations to make it a little different.

After the first year we were, as already mentioned, reduced to just two forecasters, Steve Foster and myself. Things started to change somewhat and we saw the introduction of a few weather symbols, although we were told we hadn't to move them. One evening when I arrived at the studio I was told that we would not have another presenter with us on the weather and we would control it ourselves. I thought that would mean that I could have one of those ear pieces that all the other presenters wore but they said 'No' as it would only confuse me, having to listen to all the talk and instructions from the control room. Instead I was told that I had to watch the floor manager, who would give signals indicating how much time I had left. I wasn't too happy about this as I thought it could lead to problems. Imagine, I had to adlib the forecast, turn the cube and point out the features on the map, occasionally smile towards the camera and at the same time try and keep an eye on the chap giving me hand signals.

My concern was justified. I saw most of the signals as I was presenting the weather but never saw the last one, where he counts me down through the last ten seconds. When I saw him again, he was drawing his hand across his throat. I hadn't been told about this one but I worked it out. It meant shut up, or 'cut' as they would call it in television. I wasn't sure what to do, as I had not finished the forecast, but knowing how important timing was I just stopped talking. I thought the camera had then gone onto Richard Whiteley and, feeling very upset with my performance, I threw my arms up in the air and went face down on the desk. Unfortunately I was still in vision and this went out to the nation.

I left the studios feeling very low and decided whilst driving home that I would pack in television as it wasn't worth all the stress and upset. Little did I know then that any publicity could be good for your career. The next morning whilst listening to local radio, I heard the presenter mention my name. He said, 'Did you see the weatherman Bob Rust on Calendar last night? Did he have a heart attack?' At that I thought I must be getting famous, and my disappointment was soon forgotten and I was ready for my next appearance.

Over the next few months I was given an ear piece and the presentation started to develop with more chat with the presenters and a little humour started to come into the weather. I had improved my part by wearing fancy ties and some coloured jackets and this would often provoke some comment from Richard Whiteley, which would bring a humorous response from me. From viewers' letters I realised that it was this banter which they liked best so we rather capitalised on this fact. My biggest problem was that the link to the weather was not scripted. It would just say in the script, 'Richard – ad link to weather', and I worried that I might not have an instant reply.

The usual practice was that there would be a short, filmed piece before the weather and during this I would watch Richard. A smile would come over his face as he found something funny to say to link to the weather and he would write a note on his script. I had to be on my toes and try and think what he was going to say so that I had an immediate reply. On most occasions this worked out quite well and there were only a few times that I struggled for an answer.

Occasionally things were fixed a bit. Richard would ask me before the programme what he should ask me that night and I would take advantage of these occasions. If we were in the middle of a rainy spell, I told him to ask me if it would ever stop raining. When we went live he asked me this and I quickly replied, 'It always has done in the past.' And then I would carry on with the forecast knowing that the link and response had gone well. On another occasion I persuaded him to ask if the weather would improve enough for him to cut the grass. I said, 'You should do as I do.' He asked what that was and I replied, 'Put beer on it. It doesn't stop it growing but it comes up half cut.' Although I think the quality of my forecasts was good, I believe it was this cheerful banter which made my television career successful.

Working with Alan Hardwick was somewhat different. He would often say, 'I am going to ask you so and so, in the link to the weather.' And then he would say something completely different. I am sure that the viewers noticed a baffled look on my face and this often caused a lot of laughter and amusement between Alan and myself.

Most of the presenters started to try and get a laugh with the weatherman and I can remember one occasion with the bearded Nick Powell. He linked to the weather trying to impersonate my Yorkshire accent. I responded by saying something like, 'That is Nick Powell doing an impersonation of a mouse peering through a ball of wool.'

This was then the pattern of my television appearances – brightly dressed with some humour in the forecast whilst still trying to get them right. Occasionally I was asked to dress up to do the forecast, being such things as Father Christmas, a French onion seller and even a Ninja Turtle. There were more occasions when we did the forecast out on location, appearing at agricultural shows and many other events in the region. When working with Alan Hardwick on Lunchtime Live, we seemed to be at the Ridings Centre in Wakefield nearly once every week.

A couple of amusing incidents I remember were in the summer. I went along to some nurseries to explain how hot it was in a greenhouse in sunny, summer weather. It finished up with me wearing shorts and prancing about in the automatic watering sprays. Another occasion was at Headingley when Pakistan was playing. I was positioned behind a large group of Pakistan supporters. As we came to the weather someone had been bowled out just as I was about to speak. All these supporters leapt to their feet and cheered. I quickly responded by saying I had brought my fan club. If the forecast had gone wrong I would be in for a bit of ribbing the next evening. I remember one time when this happened and my response was that I wasn't sure whether the forecast had gone wrong or the weather had gone wrong.

There were occasions when personality guests got themselves involved with the weather. Sue Pollard was a guest one lunchtime being interviewed about her role in a local pantomime. When I started to give the forecast, which was for expected rain, she came rushing over to the weather set and as I said it would rain she started, 'Oh no it won't.' And this went on right throughout the forecast!

Guest star Lenny Henry was asked what he thought of the weatherman after I had finished my forecast. He said I was OK but he thought I should have a gimmick such as a glove puppet. The producer latched on to this and the following night I had to do the forecast with a black glove puppet called Winston.

When Cilla Black was on the show, I introduced my weather by saying that things weren't too black but surprise, surprise we were expecting some rain. I don't think she was too amused as she responded by having a bit of a go at me. She wondered who had dressed me as my socks didn't match.

Throughout my time on television I was lucky to meet many personalities from all walks of life including Sir Leonard Hutton, Former Prime Minister Harold Wilson, Cliff Richards, Ken Dodd, Phil Collins, Dave Allen and so many more. The list read like Who's Who. Looking back it seems that Calendar had a top personality on the programme every evening. The surprising thing was that most of them were very pleasant people. There were just one or two who left something of a bad taste but this is not the place to mention any names. One occasion I clearly remember, was when Joe Johnson appeared the night after he had won the Embassy World Snooker Championship. He came into the Green Room and asked if I would give him a cigarette.

One evening I was told that there was a young lad from Keighley who was a bit of a weather anorak and wanted to become a weatherman when he grew up. He had been invited to appear on Calendar to be interviewed and have a go at presenting the weather. I met this young lad who was about twelve years old and also very small. I had a chat with him and gave him a few tips on how to do the presentation. He had a go but understandably he was very nervous. We were both then interviewed by Richard Whiteley, who asked me how I thought he had done – and would he make a weatherman? I said that he had done quite well but if he was to become a weatherman he would have to be much taller. Richard asked me why, and I said that if he didn't grow a lot taller he would be the last person in Yorkshire to know it was raining. Surprisingly, this young lad was Paul Hudson!

In 1989 my wife and I were invited to a television awards ceremony. Now at that time I thought all

Proud moment in 1989 when I received the award as Sony TV personality of the year.

these awards were fixed and everyone knew that they had been selected. So imagine my surprise when I arrived and found that my name was on the list of nominees for an award. My wife suggested that I should prepare a few words in the event that I had won. Still convinced that it was a fix I said that there was no need to bother. Then the real surprise when my name was announced as the winner of the Sony TV personality of the year. I was absolutely delighted and managed to find some words to say when I was presented with the trophy.

Occasionally I am asked about bad moments, and there were a few. The one I often recall was on St Swithin's Day. I was sat in my little cupboard studio and as it had rained that day I decided to start by saying, 'Good evening. Well it is St Swithin's Day and it has rained, but I can assure you that it will not rain for the next forty days.' The director soon said 'Cue Bob', and I started. Sadly I couldn't remember whose day it was. My mind was going through several saints' names and I was saying nothing. I felt terribly desperate. This was live television and I didn't know what to say. After what seemed like a lifetime, the director realised I was in trouble and came to my rescue. He shouted down the ear piece, 'Swithin's, you daft bugger!' I then carried on as though nothing had happened. Leaving the studio I had convinced myself that no one would have noticed. I was later to learn that it wasn't even St. Swithin's Day, which was the next day. When I appeared the following evening I started by saying, 'Good evening. Did you spot last night's deliberate mistake?'

With all of this, television was a very happy and enjoyable time and not really like work at all.
After several years things started to change somewhat as sponsorship entered the weather. As you then could not have sponsorship within a news programme a special weather slot was shown before Calendar. This was fine in that I had a fixed amount of time to give the forecast and we introduced computer-generated graphics, which greatly improved the presentation. The forecast was still transmitted live, but being alone in a small studio with a remotely controlled camera it was a bit difficult to put a few funnies into the presentation. However I also retained the spot on

Calendar which continued to be enjoyable with its trademark banter.

Sometime later there was some rescheduling. I was very disappointed, as despite many letters of protest from viewers, my slot in the programme was lost. Despite this I still had my own little weather programme which I started to improve. It was now shown afterwards, and I would try and pick up on an item at the end of Calendar to try and get a laugh and retain some continuity. The weather was still transmitted live with the adrenaline still flowing.

About a year later the weather got a new sponsor and further changes were introduced, the worst one being that the presentations would be pre-recorded. This was really quite a blow to me as there is nothing like live television. Granted you are likely to make a few mistakes but the viewers like them. Mistakes make you seem ordinary and the viewers feel they can identify with you. With these pre-recorded forecasts, television started to lose a lot of its appeal to me and after a couple of years I started to think about early retirement.

Being a TV personality

I was first recognised in the street and asked for an autograph after a few months of being on television. Looking back I suppose that it was quite exciting and gave a feeling of importance. However after a time it became something of a problem and suddenly you find you are no longer a private person but in a way you become public property. I never really took to this personality role and to this day I can't understand why people want autographs. Although many who work in television regard themselves as 'stars' or 'personalities', to me they are just well known and become something of a household name, like bleach. The only thing is that, if you are on television and do not get recognised, television is not working for you. Although being recognised is a problem, it is quite nice to be approached and spoken to in the street.

It is the occasions when people don't speak to you, but about you, which are embarrassing. I can remember one occasion when walking through Wetherby with my wife. Two elderly ladies were coming towards us. One of them walked right up to me, looked into my face, and then turned to her friend and said, 'Yes. It is him off the telly who reads the weather.' She never said a word to me but then walked on with her friend. Similar things happen in restaurants, when people on a nearby table will be staring at you and talking about you and this makes you feel most uncomfortable. Because of all of this, you tend to avoid public places. Visiting shops is one of the worst problems and you tend to hide away in corners and not have eye contact with people. In some shops you are aware that the salesman knows who you are but pretends that he doesn't. Whilst serving you he will start and talk about the weather, hoping you will make a proclamation and identify yourself. It is quite fun not to respond to his probing.

Even going on holiday is still a problem. Whether it be in this country or abroad there will still be people there from the region. In airports you tend to hang around in corners, hoping not to be recognised, but you can never avoid it altogether. On one occasion I went to Spain with my family. When we arrived we put on our beach wear and went to sit by the pool. The weather was beautiful with sunny skies and very hot. On the next table there was an elderly gentleman wearing a straw hat and trousers complete with braces. Even behind sunglasses I was aware that he was sneaking a look at me. Suddenly he pushed back his straw hat, turned to me and said, 'Now then mester Rust. Do you think it will stay fine for us?'

The last time I visited my beloved Doncaster Rovers, I decided that I would go just in front of the stand to avoid a lot of people seeing me. They were playing Bradford City and throughout the first half a man in the stand, obviously a Bradford supporter, was shouting abuse at the referee. Near the end of the first half he was really giving him a barracking. I turned round to see who this chap was. As I did so, he looked at me and shouted, 'Bob Rust. That referee's decisions are worse than your forecasts.' You can probably imagine the crowd around me then turned to look at me, and for the remainder of the game it was quite uncomfortable.

Having said all of this, I suppose it is understandable how people react when they see you. Going into their homes each evening they do feel that they know you. Although I retired about six years ago I still have a problem. People recognise the face but have forgotten the name or occasionally don't remember that you were on television. I am often approached by people who say, 'I know you, don't I. Do you belong to our golf club?' or something similar. Occasionally people say I look exactly like that chap who presented the weather on Calendar. Here I can have a bit of fun and say that he is my brother.

When appearing regularly on television you soon start to receive letters from viewers. Over the years I received literally hundreds. The majority of these are from very nice people who write to you about things concerning your dress or something funny you said on television. Many ask for an autographed photograph or say are that you related to someone they know who has the same name. Rather surprisingly very few letters are about the weather, although occasionally when you get it wrong, the letters do start flooding in.

You have to be very careful in your forecasts that you do not put too much emphasis on bad weather, particularly at the coast. If you forecast poor weather on the East Coast on a summer weekend, the letters will come flooding in from the seaside traders. Many years ago I had a long running battle with the mayor of a Lincolnshire resort. He said that I was always getting the forecast wrong and that the weather there was always good. It was agreed that I would go there with a camera team. Although this was in the summer, the day was terrible. It poured it down all day. I appeared like a drowned rat giving the forecast and finished up by pouring water out of my shoe.

Of the few critical letters I received, one was from a viewer complaining about my English. I had said, 'We have got high pressure sat over the country.' He was quick to point out to me that it should have been 'sitting over the country'. Even if letters were a little critical, I did personally answer all of them, except the odd one from Anon.

Quite a number of viewers sat down and wrote to me in verse, and I still retain a file of all of these poems. There are too many to include here but I will give you three examples.

This was from B. Stacey of Sutton Bridge:
If you want to know the weather
Here's a man that you can trust
Always switch to Calendar
And look into Bobby Rust
He knows when pressure is low
And also when it's high
He even shows you pictures
From a satellite in the sky
So if you are going fishing
Especially for cod
Always look into Calendar
And put your trust in Rusty Bob.

And from Marie Whiteley of Scarborough:
For me, Calendar is a must
And also is Bob Rust
He is tall, slim and sincere
So Richard Whiteley please don't sneer
Puppets, gimmicks we don't rate
It's Bob we all appreciate

I rely on Bob on how to dress
If he's not right I could look a mess
Bikini clad and raining hard
Or woolly jumper and me like lard
Oh Bob, if you only knew
How much faith we have in you
Far more than the other side
Whose predictions never ever coincide
So remember Bob, when it's hard to smile
It is really all worth while.

And finally from Frank Staples of Yeadon:
I've just seen Bob Rust on our telly
And tonight he was going full blast
For he wants us to know
'At we'll soon have more snow
An he ses when it comes – it will last.

But mi house is full of frozen pipes
An' I'm four hundredth on plumber's list
An' coalman's hoss 'as gone on strike
Nay Bob, is there nowt you can do about this.

Now you can't mess about wi nature
That is summat you just cannot do
An' that's wot's causing bad weather
But don't worry Bob, we're not blaming you!
But the biggest mistake they ever made
Wor when they changed it to Centigrade
For the weather we had was quite alright
When we used to have it in Fahrenheit.

After being on television for a while, you then start to receive letters asking you to give talks, open village fetes and a variety of other events. At first I found some excuse for not doing these as it was really something that I thought I wasn't able to do. It would seem that if you appear on television people think that you are a natural entertainer. I can assure you that giving a television weather forecast is much easier than speaking to a live audience. Eventually you succumb to one of these invitations and go along and give a talk.
Now I know that I could not read from a script, so you map out a talk in your mind and keep going over it time and time again. On the day you go along to the venue and meet the organisers over a meal, making pleasant conversation. All the time you are very nervous and you keep trying to remember your talk. Eventually you are introduced and away you go. It really can be nerve racking and occasionally your mouth becomes totally dry and you can hardly speak. You can gain a little time to recover by asking if there are any questions. It is total relief when you finish and sit down. You then listen to a vote of thanks and receive some applause. As you drive home you relive the talk and realise how it could have been so much better.

Nowadays, after having given a few hundred talks to a wide variety of groups, things are not such a strain. I prefer to speak to ladies and I find that they are the most appreciative audiences. Over the years I have spent many a happy hour with members of the WI and ladies luncheon clubs. Opening garden parties and summer fairs is not that rewarding. The organisers invite you in the belief that you will get a lot of people to come along. The usual format is that you say a few words to open the proceedings and then you are taken around to all the different stalls where you are expected to show your skills at dart throwing or knocking over skittles. After some light refreshment you are asked to judge the best cabbage in the horticultural tent or the bonny baby contest. I find I excel at drawing the raffle.

Over the years I have enjoyed many of my so-called public appearances although some problems do arise. On one occasion I was compering the Miss York Evening Press competition at a York hotel. It was a pleasant experience, interviewing a group of pretty girls. I was in the middle of one interview when the fire alarm interrupted and we had to vacate the hotel. I think that the girl I was interviewing went on to win the title.
I was asked if I would take part in a bed push in Rotherham to raise money for the local hospital.

Without checking the details I agreed to go along. It turned out that I was to be dressed in a nightgown with a silly hat and then pushed around Rotherham all day. It was a bit embarrassing but I believe that a lot of money was raised, so at least it did a lot of good.

Another interesting event I was invited to open was the centenary celebrations of the Langdale flood of 1892. It was ironic that at that time we were in the middle of a long dry period. However, on the day it absolutely poured it down and all the celebrations took place undercover. On those occasions you feel terribly sorry for the people who put so much work into organising the event.

Although most events are well organised you have to be careful what you agree to do. One Friday afternoon a lady phoned me and asked if I would attend her charity art sale and draw the raffle. She explained that Geoffrey Smith, the former television gardener, had agreed to do the job, but unfortunately he had had to pull out at the last moment. That seemed all right so I went along the next day. It turned out to be a complete flop. There were only about three people there and I spent the small fee she gave me on the only picture sold that day.

A man in Doncaster wrote to me and asked if I would come along to the town's Dome where he was organising a very big charity show. He explained that there would be numerous bands and floats and the Mayor of Doncaster would open the proceedings. In addition to all the entertainment, he had arranged an aircraft flypast from the local RAF station at Finningley. This sounded like a worthwhile event and also it was in my hometown. I wrote back and agreed to come along. A few days later, I received a telephone call from the mayor's office telling me that it was all a lot of nonsense, the mayor would not be attending and there was definitely no RAF flypast. I contacted the chap organising the event and said I was sorry that I could not now attend. Whether or not any event took place, I really do not know.

An invitation to appear at a St Valentine's Ball along with James Whale, Dave Berry and the Cruisers, and Helene Palmer who played Ida Clough in Coronation Street, turned out to be a disaster. I arrived first and was shown into the VIP lounge where I met the two chaps organising the event. They soon started to drink and then went out into the ballroom. The other celebrity guests soon arrived and after quite a long time nothing seemed to be happening. I spoke to James Whale and said I would go into the ballroom and get things started. What a surprise! The two organisers were sat on the stage, well the worse for drink and there were only about thirty people in the dance hall. I stood there and tried to crack a few jokes and then had a few words with Helene Palmer who had now appeared on the stage. Not having any direction as to what to do next I introduced James Whale. Now one of the big things for the evening was to choose a St. Valentine's Queen and two runners up. James then had the house lights turned on and selected the three nearest females as prize-winners. We gave them each a cheque, which I had been given earlier, and put a sash on the winner. James and I were ready for a quick exit, so I announced that Dave Berry would be on the stage shortly and we both then made off into the car park and drove away. What happened there for the rest of the evening, I have no idea.

Although there have been a few disasters, it has been nice to meet many of the viewers over the years.

On one occasion I agreed to take part in a charity 'bed push' in Rotherham – without checking the details. It turned out that I was to be dressed in a silly hat and nightgown!

It was a bit embarrassing but I believe that a lot of money was raised.

From anorak to weatherman! Here's a young Paul Hudson, aged about twelve, just before I appeared on Calendar to talk with Bob about my 'bizarre obsession with the weather'. The picture was taken in my back-garden weather station in Keighley, where I'm seen holding a rain gauge.

(Photograph supplied by kind permission of Telegraph & Argus, Bradford)

5 Paul Hudson - TV Forecaster

How I became a weather forecaster

I have always been obsessed with the weather, so my mum tells me. Certainly as long as I can remember I have been excited by changes in the weather, and weather extremes. When most other 'normal' children were perhaps playing football at school or hanging around the back of the bike shed, I would be religiously observing the weather in my 'Casella' weather diary, of which I had almost ten years of daily records by the time I went to university. Sad or what!

My first recollection of what was to become quite an obsession was probably during the long, hot summer of 1976 – more of which has been said earlier in the book. I was five years old, and I somehow vividly remember stifling heat one afternoon round the back of our house in Exley Crescent, Keighley, whilst my mum was reading what I think was a copy of The Sun with a headline that read something like 'Britain hotter than Spain'. The next few years were something of a blur (as they might be for someone of that age!), but at some stage when I was probably no older than six or seven, I do remember vividly what ultimately started my life in weather.

It had been another hot, summer's day when the skies suddenly became as black as night. Our garden, situated on the north-western slopes of the Worth Valley, had brilliant views of Haworth Moor to the south-west and Ilkley Moor to the north-east. By the middle of that hot, sultry afternoon the heavens opened and to this day I cannot remember a better display of lightning, grounding across Hainworth and Haworth with amazing frequency, coupled with incredible claps of thunder as the air exploded around me. I was completely soaked by the end of the thunderstorm, having sneakily managed to sit on the front door step despite being told umpteen times by mum that if I did not get inside straight away then I would get 'a right good hiding'! So I either inherited a 'weather interest gene', or through my early experiences of freak weather, I became a self-confessed 'weather anorak'.

By the time I was eight years old, I had purchased a thermometer, which was positioned in the shade on the garden hut, and made a rain gauge from an empty Fairy Liquid bottle, which was sunk into the ground. Handily enough, we had the Yorkshire Post delivered every morning, where, on page 2, could be found a wonderfully in-depth weather service. It included observations across Yorkshire and Lincolnshire and the previous day's satellite picture (which looking back on was of limited use, but fuelled my interest further). So religiously I would cut out the forecast service (much to the constant disgust of my dad who liked to read the front page of the newspaper before it was cut in two!) and stick it in a daily scrap book. This also included weather comments of my own and readings from my make shift observation site in the back garden. A daily record could not be complete without a barometer check, so, not having a barometer at this stage, I pestered Edna Shires next door EVERY day at 6pm on the dot, so I could read the barometer in her hallway. At the time I thought nothing of this but looking back I'm sure she must have told the neighbours 'that lad next door is very odd'!

By the time I was nine (I'm sure to the great relief of poor old Edna) I had my own aneroid barometer. Then I got the best Christmas present ever the following year, when my granddad made me a wonderful Stevenson screen together with max/min thermometer, psycrometer, state of the art copper rain gauge, and anemometer. And there it stood for ten years, north facing at the top of our garden in Keighley, enabling me to make entries in my Casella weather register every day at 6pm prompt! When we were away on holiday, my granddad would take the readings, and failing that we just had to come back home!

By the time I was twelve, I regularly contributed my observations to the Keighley News and appeared in several newspapers. These included the Bradford Telegraph and Argus and the Yorkshire Evening Post, who, unsurprisingly, were intrigued as to why a twelve year-old boy wasn't playing football or doing more normal things at his age! By this time I had discovered television weather forecasts, regularly watching a certain Bob Rust on YTV, who, it has to be said, was something of a hero to me. Never in my wildest dreams did I expect to follow in the footsteps of the man who has to be credited with introducing the concept of ludicrously loud jackets and ties on TV. He once told me that it was always a cunning ploy to distract the viewer at times when the forecast was more problematical than normal, so they would concentrate more on what he looked like and forget what he was saying. I did think he was pulling my leg at the time, but I now find it a very useful ploy indeed!

But I was about to get my first break on TV. One afternoon, in the middle of a maths lesson, the headmaster knocked on the door and asked to see me outside. Now you can just imagine what was going through my mind and I just couldn't think of what I could possibly have done wrong! As it turned out, YTV's Calendar programme had contacted the school to see if I would be interested in appearing on the show that evening with Richard Whiteley (who was at that time the main presenter) and Bob Rust to chat about my bizarre obsession with the weather. Well, it has to be said that I was filled with horror at the thought of appearing live on TV, and had to be persuaded by the headmaster that I would regret not doing it, and besides, I would receive a £25 appearance fee for my troubles. I immediately agreed! Just think, I could buy myself a brand new barometer for that! (Amazingly, to this day, almost twenty years later, the appearance fee remains the same.)

So I arrived at the studios, sat in the 'Green Room' and can barely remember the TV interview – I was shaking with nerves. But I got through, and such was the ordeal I made a mental note to myself never to set foot in front of camera ever again, not even for another £25!

The next few years were dominated by schoolwork, O levels, A levels and work experience. Luckily for me, the Leeds Weather Centre was opened in 1985, and I managed to secure a two-week period of work experience which finally made my mind up that I wanted to have a career as a forecaster with the Meteorological Office. At the time, the head of the Leeds office was a rather grand fellow called Dave Langley, who used to scare me to death with his military-style ways. He was always addressed as Mr Langley, never Dave, and if you dared to come to work without polished shoes then you were in for a good dressing down! (The Met Office was then, and still is, part of the Ministry of Defence. Most staff at Leeds had been transferred from the Met Office at RAF Bawtry).

But I will always be grateful to Mr Langley, who strongly advised me to work hard at school, but more importantly, concentrate on maths and physics and get a degree if I could. I later learned that many potential Met Office recruits had been given bad careers advice from school (now there's a surprise) and studied geography instead. For years this was not accepted as a relevant qualification to train as a forecaster at the Met Office college, because as stated earlier, the art of weather forecasting relies heavily in understanding the physics of the atmosphere. Moreover by the time I left university, new forecaster recruits had to have a degree.

So onto Newcastle University to study physics. Even then, my dad would continue to keep my weather records, every day, day in day out while I studied hard for what turned out to be a brilliant time of my life, but also one of very hard work – physics is not easy! In years two and three I chose to pursue geophysics and planetary physics, which contained large sections of meteorology and oceanography. My knowledge of these subjects turned out to be invaluable, and I proudly left Newcastle with a first class honours degree.

But it wasn't straight to the Met Office. To my immense disappointment at the time, I discovered that the Met Office only recruited once a year, and I had missed the cut off date! So I spent a year working for a company called Ensign Geophysics in Weybridge, Surrey, analysing sub-surface geological sections to find potential oil deposits. Now if you think making ten years of weather observations sounds tedious, you should spend a day doing this! I was almost brain dead by the end of the working day, but at least it further cemented in my mind how important it is to do something in life that you are interested in – and there was only one place where I was going!

I eventually made it into the Met Office in September 1993 after a successful interview, following which I was sent a letter. I was told that the interview panel 'saw certain characteristics of mine which led them to believe that I may at some time in the future be suitable for TV weather forecasting, and would I like to go for a screen test in London?' As it turns out, having thought at the time that I was the only person to receive such a letter, I was one of many - the competition was intense! But first and foremost I had to concentrate on my new career, which began with a six-month course at the Met Office College in Reading, set in beautiful grounds at Shinfield Park next to the European Centre for Medium-range Weather Forecasting (ECMWF).

I have to say that this was one of the best times of my life. There were probably 120 new graduates, living on site for the duration of the course and this is the best bit - on full pay, with no accommodation to pay for! We had a ball. I think the (subsidised) college bar didn't know what had hit it, and its very sad that as this book is being published, the college is moving with the Met Office to brand new headquarters in Exeter. I think most of my 1,200 colleagues within the Met Office will have many a happy memories of the college, and I'm sure a twinge of sadness at its relocation.

Following my first screen test at the BBC Weather Centre, I was sent what I later found out to be a standard letter which stated that I had 'potential'. But I wasn't too despondent, since I had joined the Met Office to be a forecaster and wasn't actually that keen to do TV. I think this was the mental scaring from my appearance on YTV when I was still at school that frightened me to death! One of the biggest misconceptions about forecasters is that TV and radio is all we do, whereas broadcasting actually makes up only a tiny fraction of Met Office business, a subject which is covered in depth elsewhere in the book.

I continued with the forecasting course attending many a party and meeting lots of very interesting people. The college itself was renowned (and still is) as an international centre for forecast teaching, and we had our fair share of international students. From Cyprus to Ghana, to South Africa across to Saudi Arabia (now that MUST be easy forecasting!) we all worked and played extremely hard and to this day still keep in touch (Kleanthis in Cyprus was a TV celebrity the last I heard!).

Towards the end of the course, fifteen of us were marshalled down to BBC South in Southampton – a vacancy had arisen for a second weather presenter and we were all to be screen tested. This turned out to be fun and much less nerve wracking than the London audition, and to my surprise the editor of the programme narrowed it down to just two – myself and a lovely lass called Sarah Wilmshurst (who now appears regularly on BBC national weather). So on the one hand they had Sarah, with a lovely clear un-accented voice, and me – a 'gobby' Yorkshireman with an accent as thick as mud. So it was probably best all round when the following week I received a phone call from the Met Office. This informed me that, since there was going to be a vacancy at BBC Leeds for a second weather presenter as well, it made sense to send me there, with Sarah going in the opposite direction to the south coast. I couldn't contain my glee!

One thing that is always drilled into you when joining the Met Office is the fact that forecasters are 'mobile grades' – often moved at short notice to any of the Met Offices around the country or overseas in order to gain more weather experience. (In fact this had for a long time put people off becoming forecasters, who until recently could be expected to move on average once every three years with all the upset and turmoil that can lead to). So having being told originally that I could be sent anywhere from St Mawgan to Sella Ness, you can imagine my delight at receiving my first posting notice to Leeds Weather Centre, commencing February 1994.

After a few weeks forecasting at Leeds, Alan Dorward, who himself had quite a following as the head of Leeds Weather Centre, retired from his job running the office and presenting the weather on BBC Look North (with Harry Gration and Judith Stamper). This left Darren Bett (now also of BBC National fame) to replace him, creating a vacancy for me to stand in for Darren.

It had to happen on Halloween. On October 31st 1994 I made my first appearance on BBC Look North in Leeds. It is safe to say that I had not slept the night before, indeed I don't think I'd slept properly for a week. Although I had made numerous radio broadcasts, it did not prepare me in anyway for the sheer nervousness of my first live TV broadcast. And I can honestly say, apart from just about remembering the then main presenter Judith Stamper suggesting it was windy enough for witches to fly unaided on their brooms, I can't remember a word I'd said. It was a feeling of terror, a feeling that at any moment my heart would explode out of my chest, and a fear that EVERYONE would be watching if I made a hash of it.

And I knew that there was possibly worst to come. Quite a few people were terrified of the then editor of Look North, John Williams, who now works for Channel 5 news. I had walked through the newsroom enough times to hear some of the most amazing dressing downs I'd ever heard in my life. On one occasion, a few months later, Darren Bett, had tried to inject a bit of humour into his forecast by wearing Dame Edna Everage's trademark glasses. It was really rather funny, but it is safe to say that Look North then was a deadly serious news programme, and humour really wasn't allowed. What followed as Darren made his way through the newsroom was enough to make a grown man cry, as the editor let rip at his 'irresponsibility' of daring to raise a smile. How things have changed (thank goodness)! Television can be a nasty business, and it was hardly surprising that I chose to leave by the back door, rather than face any of that. But the main thing was that I'd finished my first appearance, and over the next few days my confidence slowly grew and life on TV became much easier.

I stepped in for Darren for about six weeks in the year, but primarily I was a weather forecaster keen to get on, so when a job came up in the sought-after International Forecast unit at Bracknell I didn't really hesitate in applying. Again, there was fierce competition, but what perhaps tipped the balance for me was my previous job experience in the oil industry. It tuned out that one of the unit's main customers was BP and Shell who had oil platforms all round the world, and needed forecasts on a regular basis. So my year following graduation with Ensign Geophysics was not wasted after all! Having bought a house at Chellow Dene, Bradford only ten months previous, I was on my travels again, this time to headquarters at Bracknell to start one of the most interesting jobs in my career. I bought a house in Wokingham and genuinely believed that I would be there for life. Although as mentioned earlier, forecasters are very mobile and move many times, it is generally considered that once you reach the National Meteorological Centre in Bracknell, you never escape!

So I took on my latest challenge with relish, working as part of a 24 hour roster. We generated forecasts for Formula One racing teams, Nestlé's coffee plantations in Brazil, oil platforms around the world (South China sea, Falkland Islands, Offshore Nigeria), BBC Delhi (particularly the timing of the south-west monsoon), MBC (Arabia television station) and hurricane and tropical cyclone guidance. So after the first ten months I was happy, having gained promotion and building up a wealth of knowledge about world climate which to this day proves absolutely invaluable, especially when booking holidays abroad. There really are some hidden gems around the world that experience great weather when everyone else presumes otherwise!

Then, totally out of the blue, while I was studying on the Advanced Forecaster course back at Shinfield Park in Reading, I received a phone call from the Met Office. Apparently the editor back at BBC Leeds had a problem. Darren Bett, who had been main presenter in Leeds for three years, had, unsurprisingly, been promoted to do national weather broadcasts in London, and they wanted yours truly to replace him.

This time, I have to say the decision was much more difficult, not least because it actually meant demotion in the short term – TV forecasters at that time were actually at a lower grade than most other forecasters. It would have been a backward step career wise, and I would have had to leave the International unit, which was probably one of the most sought-after jobs in the forecasting division. But I was still only 26, and opportunities like this don't come up very often to broadcast in your own region. I also didn't want to suffer the 'what if' syndrome. So I agreed, and after just twelve months in November 1997 I was moving house yet again, this time to lovely Knaresborough and a job as main weather presenter at BBC Look North in Leeds.

Life on TV

Life as a 'broadcast meteorologist' has changed enormously even in my short time as a presenter, not to mention in comparison with the way it was done in Bob's day. When I first started I would be 'briefed' by the duty forecaster at Leeds Weather Centre as to the main thrust of the weather forecast. I would then sit with the other forecasters generating the weather graphics on an Apple Mac before sending them to the BBC's computer at Look North. All radio broadcasts would be done in a purpose-built studio at the weather centre, but for TV I would walk or drive to the studios twice a day.

Now I work in isolation at a new 'weather centre' based inside the BBC building in Leeds. Here I have a battery of computers containing observational information from the Met Office, latest computer forecast charts, latest guidance from the chief forecaster in Bracknell and a telephone 'hotline' to Manchester Weather Centre, the National Meteorological Centre in Bracknell and the BBC Weather Centre in London.

It is my responsibility to generate a forecast for our region and then liase with the other centres. This is crucial, as my local information and thoughts on the region's weather are exchanged with Bracknell, Manchester and London who can often have more computer forecast information from around the world. We then agree a consistent weather forecast or 'story' for the rest of the period. In tricky weather situations it can be a very lonely job. Only ten years ago, four other weather centres at Newcastle, Leeds. Nottingham and Norwich flanked the region, each staffed with numerous trained local forecasters. Now there are none and at times there is heavy reliance on my 'local' knowledge of the area in finalising a forecast. Put bluntly, the buck stops with me. It was always a running joke when Leeds Weather Centre was open that whenever a forecast went wrong, I would use the royal 'we' to say 'we thought the rain would come sooner', but now I have no such get out!

Thankfully the computer is immeasurably better at modelling the atmosphere compared with ten years ago, and despite the closure of so many weather centres it can be argued that the quality and accuracy of weather forecasts has not deteriorated. Indeed it always surprises me when I hear retired forecasters complain that the forecasts produced by TV weather presenters nowadays are not as accurate as those produced say fifteen to twenty years ago. This is nonsense. It is true that back then we did have more local forecasters and, probably more importantly, many observers on the ground monitoring the changing weather conditions every hour. This to a certain extent is now done from automatic observer sites, which will never be as accurate as a human being looking to the sky. Hence it could be argued that the very short term forecast for the next few hours may well have been on occasions more accurate. But three to five day forecasts were notoriously inaccurate back then and, despite still suffering from errors, are significantly more accurate now than they were twenty years ago. Indeed it was common for TV forecasts not to mention the outlook period at all.

More interestingly, closer examination of TV forecasts illustrates a very important point, and that is broadcasts back then were much shorter in duration. Whereas regional weather broadcasts used to be regularly around a minute in duration, now on most nights they are nearer 2 ½ minutes. This is of crucial importance. In one minute the broadcast need not mention too many details, and obviously avoid areas of the forecast that are uncertain. But nowadays, because of the need to fill 2 ½ minutes, we have to go into a great deal more forecast detail than would have been the case twenty years ago, and, as a general rule, the more detail that is attempted, the higher the risk

Interviewing Professor John Parkinson of Sheffield Hallam University, one of the world's leading solar scientists. It was one of my more accurate forecasts when thousands turned out because I'd promised a clear sunrise. I was under a lot of pressure for it to come right!

of errors. It also has to be remembered that the viewer's attention span is very short, and when the forecast is long that attention quickly moves somewhere else - often missing the crucial points of the forecast. Knocking the accuracy of weather forecasts will always be a national pasttime in this country, but all the evidence points to just one thing. We've never had it so good.

But being a broadcast meteorologist involves much more than broadcasting meteorology. I decided from an early stage that weather forecasts can become repetitive and boring and in fact days of exciting weather are normally few and far between. So I decided that it was important to be different. It is thanks to Bob that I wear daft coloured jackets and loud ties. In fact in the early days Bob gave me two of his jackets to wear on TV, in order, as he put it, to make an impact. I think he'd forgotten that at 6ft 2in he is considerably taller than I am and, frankly, they looked ridiculous. Needless to say that they cost more to alter than they did to buy new! But it worked, and six years later I am the proud owner of fifteen coloured jackets and literally hundreds of ties. In fact it always amazes me that when I meet people in town they are always most surprised that I am not wearing one of these ridiculous jackets, and I am actually dressed normally!

On TV my confidence continued to grow and the job became very enjoyable. It was only into the second week that the habit of putting my foot in it began, and my biggest weakness was identified. As Peter Levy once succinctly put it, 'You have an astonishing inability to engage your brain before speaking.' I remember on the Monday evening, the news item before the weather was about Lada cars being used as taxis in Hull. Apparently, Hull city councillors thought that this did nothing for the standing and credibility of Hull, and banned them. This was quite a serious news story, and I added as I began my weather forecast that I thought 'Lada cars were very much like weather forecasts – unreliable and without credibility'.
I can to this day remember the gasps in the gallery

and studio – which would sue first they were thinking, Lada cars or the Met Office! So I slipped out of the back door as usual, and, not really knowing much about the laws of libel in those days, went home as normal. The next day I had been in work for only thirty minutes when I received a phone call. It was from the regional sales director at Lada cars (UK). He was incandescent with rage, 'What a slur on the Lada car! We have tried to shake off the poor image of our cars for years and then you say that!' I tried my hardest to suggest it was a light-hearted comment and I was sure it wouldn't affect his sales but he was having nothing of it. And then came the crunch when I have to say I thought I'd had it. He said, 'We have thought long and hard about what we can do and have come to a decision. You are to be blacklisted from ever buying a Lada car on one our forecourts again.' And with that he hung up. I've never laughed so much in my life.
Over the years the on-air relationship with Peter Levy flourished. Peter has been in broadcasting for 25 years (it seems like 125 years), and really it turned out from a very early stage that we both had the same sense of humour. Whether it was about me being 'vertically challenged', or Peter drawing his pension and wearing a syrup, we really did have a laugh on Lunchtime Look North.

But it was obvious even at that early stage in 1998 that this was what the viewers wanted. We regularly had a pile of letters and e-mails with many of them on the same theme – the news is depressing enough, let's have a laugh – not to mention regular marriage proposals for Peter (most of them from 'rinsers'!). Students from around the region would, heaven forbid, get up in time for the 1.30pm broadcast and then no doubt go back to bed. It must be such a hard life for them!

I remember one fine summer's evening when I'd been sent out to do a broadcast outside the Dry Dock bar in Leeds. The temperature had been in the eighties and the place was packed with students rather worse for wear. Little did I know but half way through the broadcast a very attractive girl behind me, clearly seeing that I was live on air, grabbed her chest and started bouncing her rather ample sized boobs up and down in clear view of the camera. The camera man, clearly in a flap, quickly tried to tighten the picture to try and get her out of shot, resulting in camera wobble and a tighter shot of her, well, chest! All I could hear was hysterics in the studio, but I had no idea what was happening and simply finished the forecast as normal. But again, over the next few days, it was clear from the volume of letters that at times viewers seem to be more interested in cock-ups, mistakes and having a laugh than with the details of the weather forecast!

In general, outside broadcasts can be very nerve-wracking and more difficult than being in the studio. They often generate a large crowd, and communication with the presenters in the studio can be difficult due to the noise. Also, whereas in the studio I have a little monitor that can remind me of the sequence of weather graphics, 'on the road' I have to memorise the precise sequence as well as remember the forecast.

On one particular evening, I remember vividly being at the White Rose shopping centre with Harry Gration at the Christmas lights switch-on. Harry, the most experienced, and one of the nicest broadcasters I have worked with, will also verify how difficult a broadcast it was. He was presenting most of the programme from the centre and I had the relatively easy task of performing the weather in the middle of the broadcast. There must have been thousands of people in that shopping centre, and you could barely hear yourself think. We were also in separate locations, so we couldn't see each other and because of the noise couldn't hear each other either. Before I was about to go on air, I had been chatting with kids who were waiting to watch me perform live. It turned out that they were Leeds

A weather forecaster's life is never dull!

(Top) *There is a strong link between hot weather and crime, with an upsurge in burglaries as everyone is in the garden and the house windows are open. So every year I take part in a crime prevention campaign in Keighley, during which I am arrested for getting the weather wrong again!*

(Bottom) *Dressed as a town crier, I was shouting for Bradford during its campaign to become the European capital of culture.*

(© Keighley News – 2)

By far the biggest honour I have received was an invitation to be a guest at Harewood House on the occasion of the Queen's Golden Jubilee. Here I am, photographed with Mel B.

United fans, and, being a Bradford City supporter, I couldn't resist winding them up. That proved to be a bad mistake!

I started the weather forecast, kneeling down in Santa's grotto, which was covered in fake snow. I noticed to my right, out of the corner of my eye, that a fake snowball fight had begun. As if by perfect timing, as I just began my next sentence with mouth open ready to deliver my next word, one of the Leeds fans' released a snowball which flew straight into the back of my mouth. It never touched the sides. The fake snow literally bonded to my throat, and I began to choke live on air! I knelt there gasping for air, clutching my throat, and for ten seconds, unable to communicate. Moreover, with Harry unable to see or hear me, no one knew what was happening. Finally my time ran out and Harry picked up right on cue, totally oblivious to the fact that his colleague had just nearly choked to death on air! The local newspapers had a field day, and it even appeared in the national tabloids!

As I became more established, invites began for every outdoor and indoor event under the sun. Most of these are very enjoyable and I feel it is a privilege to be invited to so many, from places as far flung as Kettlewell to King's Lynn. But some don't always go to plan. I remember the very first event that I was asked to open. It was a village show in West Yorkshire and I was only six months into the job, so you can imagine my surprise and delight when I received a letter from the show's secretary requesting that I open the event. Moreover, they would pay me £50 for my time! I thought I had hit the big time.

So with much anticipation I set off, not put off by the fact that it had rained all morning and arrived about an hour too early. After much small talk with the organisers, I couldn't help noticing people were staring at me. Could it be that I was famous? Could it be my ludicrously loud yellow jacket? No – it was much simpler than that. It turned out that nobody had ever heard of me. In fact one guy came up to me and said, pointing to at least fifteen people who had turned up late, 'We're all sat in the tent and my wife's just asked me to ask you – do you work at the bank in town?' To cap it all, an old lady came up and asked me for my autograph. 'Yes,' I said, feeling rather pleased with myself, 'with pleasure.' She then added, 'What's your name again?'

A number of events require you to do a bit of judging, whether it be the best vegetable stall, best hanging basket or best kid's fancy dress. On one occasion I had been asked to do a town's Village in Bloom competition. The town will for obvious reasons remain anonymous, but I remember arriving one beautiful sunny Saturday afternoon in the Lincolnshire countryside. The event had been well publicised in the press, and I'd even managed to mention it on Look North that evening. But as I pulled up in my car, something seemed wrong. Had I got the wrong village? Had I got the wrong day? No. But the place was absolutely deserted. The only thing that was missing was the tumbleweed. No one, literally no one, had turned up. Within ten minutes the four members of the committee in charge of the event had arrived, and I began my task. It was the longest hour of my life. I couldn't wait to go home!

Even judging what would seem on the surface to be relatively straightforward events can surprise. I remember once being asked to judge a kids' fancy dress competition, only to get a mouthful of abuse from a young mother who had spent all week on her daughter's costume, only for me to give it third prize! She finished by bursting into tears, and hardly surprisingly I've not judged a similar competition!

And then there's the curse of the weather. Whatever the weather, you always have your first line when opening an outdoor event. If it is hot and sunny, you receive warm personal thanks from the crowd. As I always say to the crowd when the sun shines, 'Look what happens when you invite me!' Of course when it rains it really is entirely your fault, and the usual line would be, 'The forecast was right. It's the weather that's gone wrong!' I have to say that the organisers of the annual Honley Show really did think that I had cursed their event. After being invited early in 2000 to open what is one of the most enjoyable and friendly shows in the region, the area had unrelenting rainfall through May and June – in fact the nearby River Calder burst its banks with devastating results. The June 2000 show was consequently cancelled. The following year I was asked to open the event again, only for the foot and mouth epidemic to cause its cancellation for the second year running. So it must have been with some reluctance when they asked me back for a third time. Even the Huddersfield Examiner claimed that I had cursed the event, suggesting that I would not dare turn up if it rained! But the event on June 8th 2002 went smoothly enough, although heavy overnight rain did render parts of the showground waterlogged.

There are enormous privileges to the job. You are invited to many events whether it be country shows, top football matches or restaurant openings, and always receive first class treatment at Yorkshire County Cricket Club. But by far the biggest honour that I have received was an invitation by the Lord Lieutenant of West Yorkshire to be a guest of the Queen and the Duke of Edinburgh at Harewood House in celebration of Yorkshire Day and Her Majesty's Golden Jubilee. It was a marvellous occasion on a bright, warm and sunny July day, with a lunch of Poached Esk Valley Wild Salmon washed down with fine white wine – I remember Prince Philip seemed to particularly enjoy that bit. The Queen was breathtakingly beautiful in a peach summer outfit. It really is true about her dazzling smile, and I remember eating at my table with Prince Philip to my left and the Queen behind me and thinking how relaxed and normal they both were. At one point I'm sure Lady Harewood waved at me, and then explained to Prince Philip that I was the local weatherman on BBC TV. Perhaps she was telling him that the forecast was spot-on. Well that's my story and I'm sticking to it. Its amazing what good white wine can do!

More recently I have had the pleasure of working on the evening show with Harry and Christa Ackroyd, who joined Look North from YTV's Calendar. I'm sure at times Harry regrets being sat in the middle of us, because we really do enjoy winding each other up. It always amuses me that, despite the fact that Christa is usually on the wrong end of my wind-ups, most people tell her off for being too hard on me! I don't quite know why that is, but I suspect it is down to the fact that she has got the best filthy look in television. Indeed, I sometimes think that if they ever had a re-run of Dynasty, Christa would be brilliant as Alexis Carrington! But the best thing is that when the programme has finished and we have gone off air, Christa, Harry and myself are always laughing. Well, nearly always!

It seems that there are no limits to how much work there is out there for a humble weatherman. Regional TV seems to be the in-thing at the moment. In fact the chairman of the BBC has recently gone on record to say that the next few years will be about the expansion of regional TV and radio. He should have said 'continued expansion'! Not only does Look North continue to broadcast from Leeds, but there is now a programme from Hull for eastern parts of the region with – you know who – Peter Levy (he really must be close to being pensioned off) and Helen Fospero. I share my time between both programmes, resulting in four TV appearances in twenty minutes from two separate studios - all of them live! It is hardly surprising that my head is in a spin by 7pm.

I also have a regular slot on the new BBC1 programme called the Morning show, and recently made it onto BBC1's Have I Got News for You – unflatteringly about a parrot called Nico who it was claimed could do a better weather forecast than me! With fifteen broadcasts on Radio's Sheffield, York, Humberside and Leeds, it is an incredibly demanding and busy day. But I still love it and still regard myself as extremely privileged to have the job. And all this from an interest sparked when I saw a thunderstorm over Haworth Moor when I was six years old!

PAUL HUDSON'S : *Weather Guide*

(Clottus meterologicus or cumulus humilis)

Ever been mystified by the forecast ? Don't know your Cumulus from your Nimbus ! All this can now be simplified for you, with this quick reference chart all will be revealed. For best results this chart should be hung next to your T.V.

GLOSSARY OF WEATHER TERMS (or Hudson speak)

90% chance of rain.....................................unbroken sunshine for the day
10% chance of rain..................................don't go anywhere without a brolly
nitthering wind................................ ...bad enough to freeze your assets
Bob Rust or Icon....................old accurate forecaster from Black & White days
Indian Summer...................................2 hours without rain in late September
Warm Front...mind your own business !
Backendish..nervous sheep in the area
Chucking it down..................................I inadvertently drove into a car wash
Tomorrow will be dull...I'm on the Levy show again

EXPLANATION OF WEATHER PRESENTER BODY LANGUAGE

(Best when viewed during the Lunchtime forecast, BBC 1 @ 1.30)

Left arm caressing an area around Huddersfield..............................I'm on a promise tonight

Left and right arms raised, moving from side to side rapidly...................................Hello Mum !

Left arm, fist clenched, moving violently from South to North...........Secret sign to Peter Levy

Left arm moving aimlessly around map and right arm making short jabbing movements.......... ....my slide change button has developed a fault

Right arm makes swift West to East movement with left hand in pocket...................................Christa Ackroyd is walking this way to tweak my Barometer

Both arms slightly bent upwards with clenched fists, teeth bared and wild grimace on face, eyes bulging with right leg bent and raised off the ground, back arched forward, head inclined...strong wind imminent !

Why not visit www.wxsystems.com and spend hours (nay days) looking at the very best images of cloud formations; salivate over the nimbus, drool at the cumulus, ogle your favourite stratocumulus, make up your own Top Ten of titillating and tempting troposphere topics; discuss your theories about Evapotranspiration...... above all enjoy !!!